Environmental Pollution and Control

Environmental Pollution and Control

Julia Page

Callisto Reference

www.callistoreference.com

Callisto Reference,
118-35 Queens Blvd., Suite 400,
Forest Hills, NY 11375, USA

Visit us on the World Wide Web at:
www.callistoreference.com

ISBN: 978-1-64116-621-8 (Hardback)

Cataloging-in-Publication Data

Environmental pollution and control / Julia Page
p. cm.
Includes bibliographical references and index.
ISBN 978-1-64116-621-8
1. Pollution. 2. Pollution prevention. 3. Environmental protection. I. Page, Julia.
TD174 .E58 2021
363.73--dc23

Table of Contents

Preface

This book has been written, keeping in view that students want more practical information. Thus, my aim has been to make it as comprehensive as possible for the readers. I would like to extend my thanks to my family and co-workers for their knowledge, support and encouragement all along.

The contamination of the biological and physical components of the environment to an extent where normal environmental processes are severely affected is termed as environmental pollution. The pollutants can be chemical substances or a form of energy such as heat, noise and light. The severity of a pollutant is determined by its concentration, chemical nature and persistence. Air pollution, thermal pollution, water pollution, plastic pollution, noise pollution, soil contamination, radioactive contamination, light pollution and littering are the major forms of environmental pollution. It can be caused by various natural and anthropogenic sources such as construction, combustion, agriculture, mining and warfare. The book aims to shed light on some of the unexplored aspects of environmental pollution. Some of the diverse topics covered herein address the varied branches that fall under this category. This book, with its detailed analyses and data, will prove immensely beneficial to professionals and students involved in this area at various levels.

A brief description of the chapters is provided below for further understanding:

Chapter - What is Environmetal Pollution?

The introduction or presence of harmful substances into the environment is known as environmental pollution. Air pollution, water pollution, noise pollution, thermal pollution, etc. are a few of the various types of pollution. This is an introductory chapter which will introduce these significant types of environmental pollution.

Chapter - Air Pollution

The introduction of harmful or excessive quantities of substances such as biological molecules, gases and particulates into the Earth's atmosphere is known as air pollution. Some of its major causes are burning of fossil fuels, industries and forest fires. This chapter has been carefully written to provide an easy understanding of these causes of air pollution as well as its effects and control

Chapter - Water Pollution

Water pollution is the contamination of water bodies such as lakes, rivers and oceans due to human activities. The contaminants may include organic and inorganic substances. This chapter discusses in detail the causes and effects of water pollution as well as the ways in which it can be controlled.

Chapter - Land Pollution

The deposition of solid or liquid waste materials on land or underground is known as land pollution. It may contaminate the soil and groundwater along with threatening public health. The topics elaborated in this chapter will help in gaining a better understanding of land pollution as well as its causes.

Chapter - Noise Pollution

The propagation of noise with harmful effects on the activity of human and animal lives is known as noise pollution. A few of the types of noise pollution are occupational noise and aircraft noise pollution. This chapter has been carefully written to provide an easy understanding of these types of noise pollution as well as its control and regulation.

Chapter - Other Types of Environmental Pollution

Light pollution, radioactive contamination, visual pollution, particulate pollution, nanopollution, blackwater and drug pollution are some of the other kinds of environmental pollution. This chapter closely examines these types of environmental pollution to provide a thorough understanding of the subject.

Julia Page

1

What is Environmental Pollution?

The introduction or presence of harmful substances into the environment is known as environmental pollution. Air pollution, water pollution, noise pollution, thermal pollution, etc. are a few of the various types of pollution. This is an introductory chapter which will introduce these significant types of environmental pollution.

Pollution, also called environmental pollution is the addition of any substance (solid, liquid, or gas) or any form of energy (such as heat, sound, or radioactivity) to the environment at a rate faster than it can be dispersed, diluted, decomposed, recycled, or stored in some harmless form. The major kinds of pollution, usually classified by environment, are air pollution, water pollution, and land pollution. Modern society is also concerned about specific types of pollutants, such as noise pollution, light pollution, and plastic pollution. Pollution of all kinds can have negative effects on the environment and wildlife and often impacts human health and well-being.

Although environmental pollution can be caused by natural events such as forest fires and active volcanoes, use of the word pollution generally implies that the contaminants have an anthropogenic source—that is, a source created by human activities. Pollution has accompanied human kind ever since groups of people first congregated and remained for a long time in any one place. Indeed, ancient human settlements are frequently recognized by their wastes—shell mounds and rubble heaps, for instance. Pollution was not a serious problem as long as there was enough space available for each individual or group. However, with the establishment of permanent settlements by great numbers of people, pollution became a problem, and it has remained one ever since.

Cities of ancient times were often noxious places, fouled by human wastes and debris. Beginning about 1000 CE, the use of coal for fuel caused considerable air pollution, and the conversion of coal to coke for iron smelting beginning in the 17th century exacerbated the problem. In Europe, from the Middle Ages well into the early modern era, unsanitary urban conditions favoured the outbreak of population-decimating epidemics of disease, from plague to cholera and typhoid fever. Through the 19th century, water and air pollution and the accumulation of solid wastes were largely problems of congested urban areas. But, with the rapid spread of industrialization and the growth of the human population to unprecedented levels, pollution became a universal problem.

By the middle of the 20th century, an awareness of the need to protect air, water, and land environments from pollution had developed among the general public. In particular, the publication in 1962 of Rachel Carson's book Silent Spring focused attention on environmental damage caused by improper use of pesticides such as DDT and other persistent chemicals that accumulate in the food chain and disrupt the natural balance of ecosystems on a wide scale. In response, major pieces of environmental legislation, such as the Clean Air Act and the Clean Water Act, were passed in many countries to control and mitigate environmental pollution.

Workers steaming blast rocks covered in crude oil leaking from the Exxon Valdez, an oil tanker.

Pollution Control

The presence of environmental pollution raises the issue of pollution control. Great efforts are made to limit the release of harmful substances into the environment through air pollution control, wastewater treatment, solid-waste management, hazardous-waste management, and recycling. Unfortunately, attempts at pollution control are often surpassed by the scale of the problem, especially in less-developed countries. Noxious levels of air pollution are common in many large cities, where particulates and gases from transportation, heating, and manufacturing accumulate and linger. The problem of plastic pollution on land and in the oceans has only grown as the use of single-use plastics has burgeoned worldwide. In addition, greenhouse gas emissions, such as methane and carbon dioxide, continue to drive global warming and pose a great threat to biodiversity and public health.

Plastic pollution, including many single-use plastics like bags and food packaging, in the ocean.

Pollutant

A pollutant is a substance or energy introduced into the environment that has undesired effects, or adversely affects the usefulness of a resource. A pollutant may cause long- or short-term damage by changing the growth rate of plant or animal species, or by interfering with human amenities, comfort, health, or property values. Some pollutants are biodegradable and therefore will not persist in the environment in the long term. However, the degradation products of some pollutants are themselves polluting such as the products DDE and DDD produced from degradation of DDT.

Different Types of Pollutants in Nature

Stock Pollutants

Pollutants, towards which the environment has low absorptive capacity are called stock pollutants. (e.g. persistent organic pollutants such as PCBs, non-biodegradable plastics, and heavy metals). Stock pollutants accumulate in the environment over time. The damage they cause increases as more pollutant is emitted, and persists as the pollutant accumulates. Stock pollutants can create a burden for future generations, by passing on damage that persists well after the 'benefits' received from incurring that damage; have been forgotten.

Fund Pollutants

Fund pollutants are those for which the environment has moderate absorptive capacity. Fund pollutants do not cause damage to the environment unless the emission rate exceeds the receiving environment's absorptive capacity (e.g. carbon dioxide, which is absorbed by plants and oceans). Fund pollutants are not destroyed, but rather converted into less harmful substances, or diluted/ dispersed to non-harmful concentrations.

Notable Pollutants

Notable pollutants include the following groups or compounds:

- Mercury Hg.
- Persistent organic pollutants POPs.
- Ozone.
- Particulate matter PM.
- Environmental Persistent Pharmaceutical Pollutants EPPP.
- Polycyclic aromatic hydrocarbons PAHs.
- Volatile organic compounds VOCs.

Light Pollutant

Light pollution is the impact that anthropogenic light has on the visibility of the night sky. It also encompasses ecological light pollution which describes the effect of artificial light on individual organisms and on the structure of ecosystems as a whole. Primary Pollutants Pollutants which are emmited directly to the environment ard called primary pollutants.

The night sky viewed from Luhasoo bog, Estonia with light pollution in background.

Zones of Influence

Pollutants can also be defined by their zones of influence, both horizontally and vertically.

Horizontal Zone

The horizontal zone refers to the area that is damaged by a pollutant. Local pollutants cause damage near the emission source. Regional pollutants cause damage further from the emission source.

Vertical Zone

The vertical zone refers to whether the damage is ground-level or atmospheric. Surface pollutants cause damage by accumulating near the Earth's surface. Global pollutants cause damage by concentrating in the atmosphere.

Types of Pollution

Pollution is of many kinds, but the commonly known are air, land, and water pollution. Below are the different types of pollution and their distinctive comprehensive explanations.

- Air pollution

  Air is the most polluted environmental resource. It is the introduction of harmful substances in the air that results in detrimental impacts to the environmental and humanity. Air pollution reduces air quality by making it unclean or contaminated.

  It occurs when harmful substances such as foreign gases, odours, dust, or fumes are released in the air at levels that can harm the comfort or health of animals and humans, or even destroy plant life. Air pollution results from both human and natural activities.

  It is caused by emissions from manufacturing industries and power plants, vehicular emissions, smoking, natural events such as volcanic eruptions and wildfire, and burning of waste materials such as wood, rubber and plastics. The common air pollutants include hydrocarbons, volatile organic compounds (VOCs), dust particles, carbon monoxide, sulfur oxides, particulate matter, chlorofluorocarbons (CFCs), and nitrogen oxides.

- Water pollution

  Water pollution is the act of contaminating water bodies namely rivers, oceans, lakes, streams, aquifers, and groundwater. It occurs when foreign harmful materials like chemicals, waste matter, or contaminated substances are directly or indirectly discharged into water bodies. Any alterations in the chemical, physical, or biological water properties qualify as water pollution.

  Very often, the primary contributors to water pollution are human activities since they introduce substances that contaminate the water with harmful chemicals and toxic materials. Water pollution is categorized into point source, non-point source, and groundwater. Point source water pollution occurs when the contaminants enter a water body from a single identifiable source while non-point source occurs as a result of cumulative effects of different amounts of contaminants.

  Groundwater pollution occurs through infiltration and affects groundwater sources such as wells or aquifers. Water is the considered the second most polluted environmental resource after air pollution.

The causes of water pollution include:

- The discharge of industrial waste in water bodies.
- Agricultural pollution: practices such as the use of pesticides and other agrochemicals that get washed into water systems or infiltrate into groundwater.
- Discharge and dumping of waste materials and home made products such as plastics and detergents into water bodies.
- Oil spills.
- Natural disasters such as floods and hurricanes that lead to intermixing of water with dangerous substances on the land.
- Sediments from soil erosion.

- Land Pollution

  Land pollution is the destruction or decline in quality of the earth's land surfaces in term of use, landscape and ability to support life forms. Many times, it is directly and indirectly caused by human activities and abuse of land resources. Land pollution takes place when waste and garbage is not disposed off in the right manner and as such, introduces toxins and chemicals on land.

  It also occurs when people dump chemical products to soils in the form of herbicides, fertilizers, pesticides, or any other form of the consumer by-products. Mineral exploitation equally leads to the decline in quality of the earth's land surfaces. In this regard, it has grave consequences for human health, plant life, and soil quality. Acid rain, construction sites, solid waste, mineral exploitation, agricultural chemicals, and deforestations are the primary causes of land pollution.

- Soil Pollution

  Soil pollution takes place when chemical pollutants contaminate the soil or degraded by acts such as mining, clearance of vegetation cover, or topsoil erosion. Usually, it happens when human activities directly or indirectly introduce destructive chemicals, substances, or objects into the soil in a way that causes damage to the immediate earthly environment.

  As a consequence, soil losses its value of natural minerals and nutrients compositions. Soil degradation also contributes to soil pollution, and it occurs as a result of over-grazing, over-farming, or mining activities. The notable causes of soil pollution include agricultural farming activities, waste dumping on land, industrial activities, mining, and acid rain. Lose of wildlife habitat, soil erosion, reduced crop yields, contamination of food, and desertification are the precise consequences of soil pollution.

- Noise Pollution

  Noise pollution is mostly an undesirable sound or sound which generates horrible discomfort on the ears. It is measured in decibels (dB) and sound levels beyond 100 dB can cause permanent hearing loss. The industrial sound limit according to the World Health Organization (WHO) is 75 dB. In the contemporary society, noise has become a permanent aspect owing to the daily activities such as transportation including airports, traffic and railroads, industrial manufacturing, construction works, and concerts.

In contrast to the other types of pollution, noise pollution lacks the element of accumulation in the environment. It merely occurs when sounds waves of intense pressure reach the human ears and may even affect the body muscles due to sound vibrations. Noise pollution similarly affects marine and wildlife animals in the same manner it affects humans, and can even cause their death.

- Thermal Pollution

  Thermal pollution occurs when water bodies are degraded in terms of altering their temperatures. It commonly happens when people or industries undertake activities that suddenly decrease or increase the temperature of a natural water body which may include lakes, rivers, oceans or ponds.

  Thermal pollution is currently a huge menace and is mainly influenced by power plants and industrial manufacturers that use water as a coolant. Urban stormwater runoff from parking lots and roads also discharges water of elevated temperatures into adjacent water bodies.

  When water is either used as a coolant, discharged from stormwater runoff at elevated temperatures, or released from reservoirs with unnaturally cold temperatures, it changes the natural temperature of water bodies. Therefore, thermal pollution is one aspect of the wider subject of water pollution. The alterations of natural water resource temperatures can have dire consequences on aquatic life and the local ecosystems.

- Industrial Pollution

  Industrial pollution is the release of wastes and pollutants generated by industrial activities into the natural environment including air, water, and land. The pollutants and wastes from industries encompass air emissions, deposit of used water into water resources, landfill disposal, and injection of toxic materials underground.

  Industrial pollution can adversely damage plants, kill animals, cause ecosystem imbalance, and degrade the quality of life. Leading industries such as power plants, steel mills, sewage treatment plants, heating plants, and glass smelting among other production, processing and manufacturing companies are the contributors to industrial pollution.

They release smoke, effluents, material wastes, toxic byproducts, contaminated residues, and chemical consumer products that eventually end up in the environment thereby causing pollution.

- Light Pollution

  Light pollution occurs due to lengthened and excessive use of artificial lights, such that it results in the brightening of the skies. As a consequence, it upsets the activities and natural cycles of wildlife and also affects the welfare of humans. Whenever artificial lights are used where they are not intended, it causes a nuisance.

  For instance, too much outdoor light intruding into neighbor's bedrooms can disrupt their sleep. Likewise, too much indoor light has implications on the health of the inhabitants of that room. Light pollution is also referred to as luminous pollution or photo pollution. The types of light pollution include glare, light trespass, and sky glow. City lighting, advertising and billboards, and nighttime entertainments are some of the main contributors of light pollution.

- Radioactive Pollution

  It is very uncommon for radioactive pollution to occur, but when it happens the outcomes are catastrophic to the environment and to human health. Due to its intensity, it can cause massive human death, cancers, serious lung disease, organ damage, birth defects, and grave destruction to plant life and animals.

  It can also destroy soil, and cause water and air pollution. In the event that there is a radioactive pollution, it can take many years for the environment to regain its natural capacity. The sources of radioactive pollution include uranium mining operations, nuclear power plant accidents, and inappropriate nuclear waste disposal.

2

Air Pollution

The introduction of harmful or excessive quantities of substances such as biological molecules, gases and particulates into the Earth's atmosphere is known as air pollution. Some of its major causes are burning of fossil fuels, industries and forest fires. This chapter has been carefully written to provide an easy understanding of these causes of air pollution as well as its effects and control.

Air pollution is a change in the physical, chemical and biological characteristic of air that causes adverse effects on humans and other organisms.

Major Air Pollutants

Criteria Pollutants

Clean, dry air consists primarily of nitrogen and oxygen—78 percent and 21 percent respectively, by volume. The remaining 1 percent is a mixture of other gases, mostly argon (0.9 percent), along with trace (very small) amounts of carbon dioxide, methane, hydrogen, helium, and more. Water vapour is also a normal, though quite variable, component of the atmosphere, normally ranging from 0.01 to 4 percent by volume; under very humid conditions the moisture content of air may be as high as 5 percent.

The gaseous air pollutants of primary concern in urban settings include sulfur dioxide, nitrogen dioxide, and carbon monoxide; these are emitted directly into the air from fossil fuels such as fuel oil, gasoline, and natural gas that are burned in power plants, automobiles, and other combustion sources. Ozone (a key component of smog) is also a gaseous pollutant; it forms in the atmosphere via complex chemical reactions occurring between nitrogen dioxide and various volatile organic compounds (e.g., gasoline vapours).

Airborne suspensions of extremely small solid or liquid particles called "particulates" (e.g., soot, dust, smokes, fumes, mists), especially those less than 10 micrometres (μm; millionths of a metre) in size, are significant air pollutants because of their very harmful effects on human health. They are emitted by various industrial processes, coal- or oil-burning power plants, residential heating systems, and automobiles. Lead fumes (airborne particulates less than 0.5 μm in size) are particularly toxic and are an important pollutant of many diesel fuels.

The six major air pollutants listed above have been designated by the U.S. Environmental Protection Agency (EPA) as "criteria" pollutants—criteria meaning that the concentrations of these pollutants in the atmosphere are useful as indicators of overall air quality. The sources, acceptable concentrations, and effects of the criteria pollutants are summarized in the table.

Table: Criteria Air Pollutants.

Pollutant	Common sources	Maximum acceptable concentration in the atmosphere	Environmental risks	Human health risks
Carbon monoxide (CO)	Automobile emissions, fires, industrial processes	35 ppm (1-hour period); 9 ppm (8-hour period)	Contributes to smog formation	Exacerbates symptoms of heart disease, such as chest pain; may cause vision problems and reduce physical and mental capabilities in healthy people
Nitrogen oxides (NO and NO_2)	Automobile emissions, electricity generation, industrial processes	0.053 ppm (1-year period)	Damage to foliage; contributes to smog formation	Inflammation and irritation of breathing passages
Sulfur dioxide (SO_2)	Electricity generation, fossil-fuel combustion, industrial processes, automobile emissions	0.03 ppm (1-year period); 0.14 ppm (24-hour period)	Major cause of haze; contributes to acid rain formation, which subsequently damages foliage, buildings, and monuments; reacts to form particulate matter	Breathing difficulties, particularly for people with asthma and heart disease
Ozone (O_3)	Nitrogen oxides (NO_x) and volatile organic compounds (VOCs) from industrial and automobile emissions, gasoline vapours, chemical solvents, and electrical utilities	0.075 ppm (8-hour period)	Interferes with the ability of certain plants to respire, leading to increased susceptibility to other environmental stressors (e.g., disease, harsh weather)	Reduced lung function; irritation and inflammation of breathing passages
Particulate matter	Sources of primary particles include fires, smokestacks, construction sites, and unpaved roads; sources of secondary particles include reactions between gaseous chemicals emitted by power plants and automobiles	150 μg/m³ (24-hour period for particles <10 μm); 35 μg/m³ (24-hour period for particles <2.5 μm)	Contributes to formation of haze as well as acid rain, which changes the pH balance of waterways and damages foliage, buildings, and monuments	Irritation of breathing passages, aggravation of asthma, irregular heartbeat

Except for lead, criteria pollutants are emitted in industrialized countries at very high rates, typically measured in millions of tons per year. All except ozone are discharged directly into the atmosphere from a wide variety of sources. They are regulated primarily by establishing ambient air quality standards, which are maximum acceptable concentrations of each criteria pollutant in the atmosphere, regardless of its origin. The six criteria pollutants are described in turn below.

Fine Particulates

Very small fragments of solid materials or liquid droplets suspended in air are called particulates. Except for airborne lead, which is treated as a separate category, they are characterized on the basis of size and phase (i.e., solid or liquid) rather than by chemical composition. For example, solid particulates between roughly 1 and 100 μm in diameter are called dust particles, whereas airborne solids less than 1 μm in diameter are called fumes.

The particulates of most concern with regard to their effects on human health are solids less than 10 μm in diameter, because they can be inhaled deep into the lungs and become trapped in the lower respiratory system. Certain particulates, such as asbestos fibres, are known carcinogens (cancer-causing agents), and many carbonaceous particulates—e.g., soot—are suspected of being carcinogenic. Major sources of particulate emissions include fossil-fuel power plants, manufacturing processes, fossil-fuel residential heating systems, and gasoline-powered vehicles.

Carbon Monoxide

Carbon monoxide is an odourless, invisible gas formed as a result of incomplete combustion. It is the most abundant of the criteria pollutants. Gasoline-powered highway vehicles are the primary source, although residential heating systems and certain industrial processes also emit significant amounts of this gas. Power plants emit relatively little carbon monoxide because they are carefully designed and operated to maximize combustion efficiency. Exposure to carbon monoxide can be acutely harmful since it readily displaces oxygen in the bloodstream, leading to asphyxiation at high enough concentrations and exposure times.

Sulfur Dioxide

A colourless gas with a sharp, choking odour, sulfur dioxide is formed during the combustion of coal or oil that contains sulfur as an impurity. Most sulfur dioxide emissions come from power-generating plants; very little comes from mobile sources. This pungent gas can cause eye and throat irritation and harm lung tissue when inhaled.

Branches from a tree in Germany's Black Forest show needle loss and yellowed boughs caused by acid rain.

Sulfur dioxide also reacts with oxygen and water vapour in the air, forming a mist of sulfuric acid that reaches the ground as a component of acid rain. Acid rain is believed to have harmed or destroyed fish and plant life in many thousands of lakes and streams in parts of Europe, the northeastern United States, southeastern Canada, and parts of China. It also causes corrosion of metals and deterioration of the exposed surfaces of buildings and public monuments.

Nitrogen Dioxide

Of the several forms of nitrogen oxides, nitrogen dioxide—a pungent, irritating gas—is of most concern. It is known to cause pulmonary edema, an accumulation of excessive fluid in the lungs. Nitrogen dioxide also reacts in the atmosphere to form nitric acid, contributing to the problem of acid rain. In addition, nitrogen dioxide plays a role in the formation of photochemical smog, a reddish brown haze that often is seen in many urban areas and that is created by sunlight-promoted reactions in the lower atmosphere.

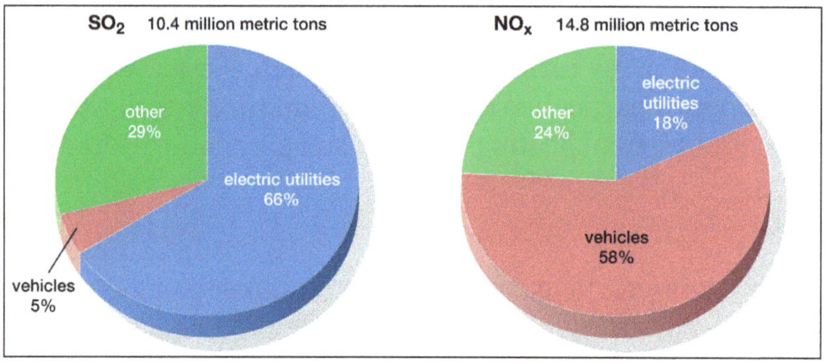

U.S. SO_2 and NO_x emissions, 2008.

Nitrogen oxides are formed when combustion temperatures are high enough to cause molecular nitrogen in the air to react with oxygen. Stationary sources such as coal-burning power plants are major contributors of this pollutant, although gasoline engines and other mobile sources are also significant.

Ozone

A key component of photochemical smog, ozone is formed by a complex reaction between nitrogen dioxide and hydrocarbons in the presence of sunlight. It is considered to be a criteria pollutant in the troposphere—the lowermost layer of the atmosphere—but not in the upper atmosphere, where it occurs naturally and serves to block harmful ultraviolet rays from the Sun. Because nitrogen dioxide and hydrocarbons are emitted in significant quantities by motor vehicles, photochemical smog is common in cities such as Los Angeles, where sunshine is ample and highway traffic is heavy. Certain geographic features, such as mountains that impede air movement, and weather conditions, such as temperature inversions in the troposphere, contribute to the trapping of air pollutants and the formation of photochemical smog.

Lead

Inhaled lead particulates in the form of fumes and dusts are particularly harmful to children, in whom even slightly elevated levels of lead in the blood can cause learning disabilities, seizures, or even death. Sources of airborne lead particulates include oil refining, smelting, and other industrial activities. In the past, combustion of gasoline containing a lead-based antiknock additive called tetraethyl lead was a major source of lead particulates. In many countries there is now a complete ban on the use of lead in gasoline. In the United States, lead concentrations in outdoor air decreased more than 90 percent after the use of leaded gasoline was restricted in the mid-1970s and then completely banned in 1996.

Air Toxics

Hundreds of specific substances are considered hazardous when present in trace amounts in the air. These pollutants are called air toxics. Many of them cause genetic mutations or cancer; some cause other types of health problems, such as adverse effects on brain tissue or fetal development. Although the total emissions and the number of sources of air toxics are small compared with those for criteria pollutants, these pollutants can pose an immediate health risk to exposed individuals and can cause other environmental problems.

Most air toxics are organic chemicals, comprising molecules that contain carbon, hydrogen, and other atoms. Many are volatile organic compounds (VOCs), organic compounds that readily evaporate. VOCs include pure hydrocarbons, partially oxidized hydrocarbons, and organic compounds containing chlorine, sulfur, or nitrogen. They are widely used as fuels (e.g., propane and gasoline), as paint thinners and solvents, and in the production of plastics. In addition to contributing to air toxicity and urban smog, some VOC emissions act as greenhouse gases and, in so doing, may be a cause of global warming. Some other air toxics are metals or compounds of metals—for example, mercury, arsenic and cadmium.

In many countries, standards have been set to control industrial emissions of several air toxics. The first hazardous air pollutants regulated in the United States (outside the workplace environment) were arsenic, asbestos, benzene, beryllium, coke oven emissions, mercury, radionuclides (radioactive isotopes), and vinyl chloride. In 1990 this short list was expanded to include 189 substances. By the end of the 1990s, specific emission control standards were required in the United States for "major sources"—those that release more than 10 tons per year of any of these materials or more than 25 tons per year of any combination of them. Important air toxics, their sources, and their environmental effects are summarized in the table.

Table: Important hazardous air pollutants (air toxics).

Air toxic	Common sources	Human health risks
Arsenic	Volcanoes, weathering of rocks and minerals containing arsenic, industrial processes such as metal smelting, burning wood that has been treated with arsenic compounds.	Acute effects include gastrointestinal disorders, headaches, hypotension, shock, and death; chronic effects include dermatitis, irritation of mucous membranes, gastrointestinal disorders, and lung cancer.
Asbestos	Natural deposits, building materials.	Chronic effects include asbestosis (lung disease), pulmonary hypertension, lung cancer, and mesothelioma.
Benzene	Burning coal and oil, vehicle exhaust, industrial solvents, tobacco smoke.	Acute effects include dizziness, drowsiness, unconsciousness, and irritation of skin and mucous membranes; chronic effects include blood disorders including aplastic anemia, adverse effects on fetal development, and leukemia.

Beryllium compounds	Burning coal and oil, tobacco smoke, natural occurrence in soil.	Acute effects include inflammation and swelling of the lungs; chronic effects include chronic beryllium disease (development of noncancerous lesions in the lungs).
Cadmium compounds	Burning fossil fuels, incineration of waste, smelting of zinc, lead, and copper.	Acute effects include lung irritation; chronic effects include various kidney disorders, sperm-count reduction, and lung cancer.
Chlorine	Papermaking, drinking water, swimming pools.	Acute and chronic effects include irritation of mucous membranes and respiratory tract.
Coke oven emissions	Coke ovens.	Chronic effects include conjunctivitis, dermatitis, lesions of respiratory and digestive systems, and various cancers.
Cyanide compounds	Automobile exhaust, chemical processing, waste incineration.	Acute effects include headache, nausea, eye and skin irritation, and death; chronic effects include central nervous system disorders and thyroid disorders.
Formaldehyde	Building materials and home furnishings, power plants, incineration of waste, automobile exhaust, tobacco smoke.	Acute effects include irritation of eyes and respiratory tract and inflammation of mouth, esophagus, and stomach if ingested; chronic effects include lesions in respiratory tract, menstrual disorders, and lung and nasopharyngeal cancer.
Lead compounds	Battery manufacturing, paints and ceramic glazes, metal products.	Acute effects include gastrointestinal disorders and death; chronic effects include anemia, neurological disorders, reduced sperm count, and fetal effects including low birth weight and mental impairment.
Mercury compounds	Paint manufactured prior to 1990–91, batteries, industrial processes, dental amalgam fillings, fish products.	Acute effects include central nervous system effects including irritability, slowed sensory and motor nerve function, blindness, and deafness (depending on compound); chronic effects include a variety of central nervous system effects including erethism, insomnia, deafness, and blurred vision and kidney damage.
Nickel compounds	Nickel processing, fossil-fuel burning, sewage sludge incineration.	Acute effects include lung and kidney damage (anecdotal reports) and gastrointestinal disorders; chronic effects include dermatitis and asthma.
Radionucleides (radon, radium, uranium)	Ambient air (radon, uranium), drinking water (uranium), uranium mines and processing facilities, phosphate fertilizers.	Chronic effects include lung, bone, and nasal cancer.
Selenium compounds	Drinking water and ambient air, some plastics and paints, pharmaceutical production.	Acute effects include irritation of the respiratory and gastrointestinal tracts; chronic effects include loss of hair and tooth decay and discoloration.
vinyl chloride	manufacture of polyvinyl chloride (PVC), which is used in pipes, furniture, and automobiles	acute effects imclude central nervous system issues including dizziness, drowsiness, headaches, loss of consciousness, and inhibition of blood clotting; chronic effects include, rarely, vinyl chloride disease, in which changes occur in the bones and skin of the fingers and sensitivity to cold is increased, and liver cancer.

Air toxics may be released in sudden and catastrophic accidents rather than steadily and gradually from many sources. For example, in the Bhopal disaster of 1984, an accidental release of methyl isocyanate at a pesticide factory in Bhopal, Madhya Pradesh state, India, immediately killed at least 3,000 people, eventually caused the deaths of an estimated 15,000 to 25,000 people over the following quarter-century, and injured hundreds of thousands more. The risk of accidental release of very hazardous substances into the air is generally higher for people living in industrialized urban areas. Hundreds of such incidents occur each year, though none has been as severe as the Bhopal event.

Other than in cases of occupational exposure or accidental release, health threats from air toxics are greatest for people who live near large industrial facilities or in congested and polluted urban areas. Most major sources of air toxics are so-called point sources—that is, they have a specific location. Point sources include chemical plants, steel mills, oil refineries, and municipal waste incinerators. Hazardous air pollutants may be released when equipment leaks or when material is transferred, or they may be emitted from smokestacks. Municipal waste incinerators, for example, can emit hazardous levels of dioxins, formaldehyde, and other organic substances, as well as metals such as arsenic, beryllium, lead, and mercury. Nevertheless, proper combustion along with appropriate air pollution control devices can reduce emissions of these substances to acceptable levels.

Air pollution from factory smokestacks.

Hazardous air pollutants also come from "area" sources, which are many smaller sources that release pollutants into the outdoor air in a defined area. Such sources include commercial dry-cleaning facilities, gasoline stations, small metal-plating operations, and woodstoves. Emission of air toxics from area sources are also regulated under some circumstances.

Small area sources account for about 25 percent of all emissions of air toxics. Major point sources account for another 20 percent. The rest—more than half of hazardous air-pollutant emissions—come from motor vehicles. For example, benzene, a component of gasoline, is released as unburned fuel or as fuel vapours, and formaldehyde is one of the by-products of incomplete combustion. Newer cars, however, have emission control devices that significantly reduce the release of air toxics.

Greenhouse Gases

Global warming is recognized by almost all atmospheric scientists as a significant environmental problem caused by an increase in levels of certain trace gases in the atmosphere since the beginning of the Industrial Revolution in the mid-18th century. These gases, collectively called greenhouse gases, include carbon dioxide, organic chemicals called chlorofluorocarbons (CFCs), methane, nitrous oxide, ozone, and many others. Carbon dioxide, although not the most potent of the greenhouse gases, is the most important because of the huge volumes emitted into the air by combustion of fossil fuels (e.g., gasoline, oil, coal).

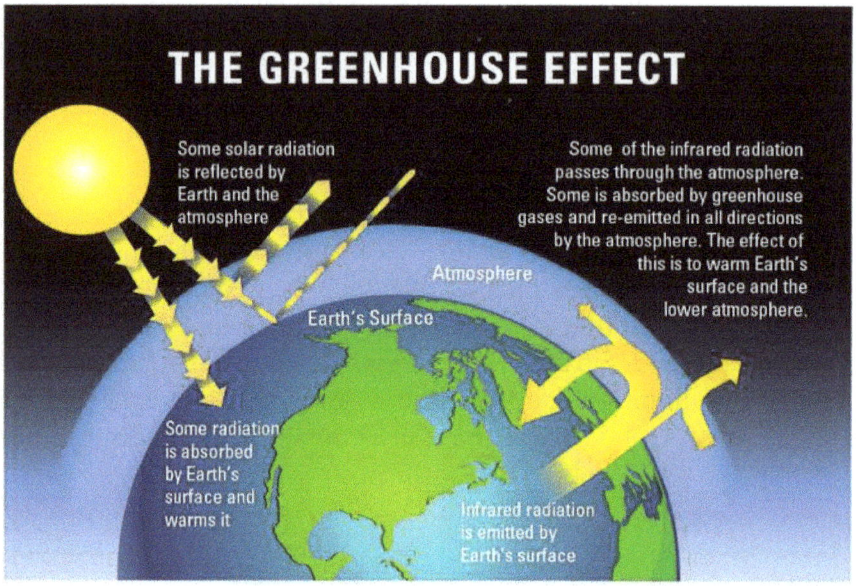

In the given figure greenhouse effect on Earth. Some incoming sunlight is reflected by Earth's atmosphere and surface, but most is absorbed by the surface, which is warmed. Infrared (IR) radiation is then emitted from the surface. Some IR radiation escapes to space, but some is absorbed by the atmosphere's greenhouse gases (especially water vapour, carbon dioxide, and methane) and reradiated in all directions, some to space and some back toward the surface, where it further warms the surface and the lower atmosphere.

Carbon dioxide is considered a normal component of the atmosphere, and before the Industrial Revolution the average levels of this gas were about 280 parts per million (ppm). By the early 21st century, the levels of carbon dioxide reached 405 ppm, and they continue to increase at a rate of almost 3 ppm per year. Many scientists think that carbon dioxide should be regulated as a pollutant—a position taken by the EPA in 2009 in a ruling that such regulations could be promulgated. International cooperation and agreements, such as the Paris Agreement of 2015, would be necessary to reduce carbon dioxide emissions worldwide.

Air Pollution and Air Movement

Local air quality typically varies over time because of the effect of weather patterns. For example, air pollutants are diluted and dispersed in a horizontal direction by prevailing winds, and they are dispersed in a vertical direction by atmospheric instability. Unstable atmospheric conditions occur when air masses move naturally in a vertical direction, thereby mixing and dispersing pollutants. When there is little or no vertical movement of air (stable conditions), pollutants can accumulate near the ground and cause temporary but acute episodes of air pollution. With regard to air quality, unstable atmospheric conditions are preferable to stable conditions.

The degree of atmospheric instability depends on the temperature gradient (i.e., the rate at which air temperature changes with altitude). In the troposphere (the lowest layer of the atmosphere, where most weather occurs), air temperatures normally decrease as altitude increases; the faster the rate of decrease, the more unstable the atmosphere. Under certain conditions, however, a temporary "temperature inversion" may occur, during which time the air temperature increases with increasing altitude, and the atmosphere is very stable. Temperature inversions prevent the upward mixing and dispersion of pollutants and are the major cause of air pollution episodes. Certain geographic conditions exacerbate the effect of inversions. For example, Los Angeles, situated on a plain on the Pacific coast of California and surrounded by mountains that block horizontal air motion, is particularly susceptible to the stagnation effects of inversions—hence the infamous Los Angeles smog. On the opposite coast of North America another metropolis, New York City, produces greater quantities of pollutants than does Los Angeles but has been spared major air pollution disasters—only because of favourable climatic and geographic circumstances. During the mid-20th century, governmental efforts to reduce air pollution increased substantially after several major inversions, such as the Great Smog of London, a weeklong air pollution episode in London in 1952 that was directly blamed for more than 4,000 deaths.

Global Reach of Air Pollution

Because some air pollutants persist in the atmosphere and are carried long distances by winds, air pollution transcends local, regional, and continental boundaries, and it also may have an effect on global climate and weather. For example, acid rain has gained worldwide attention since the 1970s as a regional and even continental problem. Acid rain occurs when sulfur dioxide and nitrogen oxides from the burning of fossil fuels combine with water vapour in the atmosphere, forming sulfuric acid and nitric acid mists. The resulting acidic precipitation is damaging to water, forest, and soil resources. It has caused the disappearance of fish from many lakes in the Adirondack Mountains of North America, the widespread death of forests in mountains of Europe, and damage to tree growth in the United States and Canada. Acid rain can also corrode building materials and be hazardous to human health. These problems are not contained by political boundaries. Emissions from the burning of fossil fuels in the middle sections of the United States and Canada are precipitated as acid rain in the eastern regions of those countries, and acid rain in Norway comes largely from industrial areas in Great Britain and continental Europe. The international scope of the problem has led to the signing of international agreements on the limitation of sulfur and nitrogen oxide emissions.

Another global problem caused by air pollution is the ozone depletion in the stratosphere. At ground level (i.e., in the troposphere), ozone is a pollutant, but at altitudes above 12 km (7 miles) it plays a crucial role in absorbing and thereby blocking ultraviolet radiation (UV) from the Sun before it reaches the ground. Exposure to UV radiation has been linked to skin cancer and other health problems. In 1985 it was discovered that a large "ozone hole," an ozone-depleted region, is present every year between August and November over the continent of Antarctica. The size of this hole is increased by the presence in the atmosphere of chlorofluorocarbons (CFCs); these emanate from aerosol spray cans, refrigerators, industrial solvents, and other sources and are transported to Antarctica by atmospheric circulation. It had already been demonstrated in the mid-1970s that CFCs posed a threat to the global ozonosphere, and in 1978 the use of CFCs as propellants in aerosol cans was banned in the United States. Their use was subsequently restricted in several other countries. In 1987 representatives from more than 45 countries signed the Montreal Protocol, agreeing to place severe limitations on the production of CFCs.

Ozone depletion, Antarctic ozone hole
NASA/Goddard Space Flight Center.

One of the most significant effects of air pollution is on climate change, particularly global warming. As a result of the growing worldwide consumption of fossil fuels, carbon dioxide levels in the atmosphere have increased steadily since 1900, and the rate of increase is accelerating. It has been estimated that if carbon dioxide levels are not reduced, average global air temperatures may rise another 4 °C (7.2 °F) by the end of the 21st century. Such a warming trend might cause melting of the polar ice caps, rising of the sea level, and flooding of the coastal areas of the world. Changes in precipitation patterns caused by global warming might have adverse effects on agriculture and forest ecosystems, and higher temperatures and humidity might increase the incidence of disease in humans and animals in some parts of the world. Implementation of international agreements on reducing greenhouse gases are required to protect global air quality and to mitigate the effects of global warming.

Indoor Air Pollution

Health risks related to indoor air pollution have become an issue of concern because people generally spend most of their time indoors at home and at work. The problem has been exacerbated by well-meaning efforts to lower air-exchange rates in buildings in order to conserve energy; these efforts unfortunately allow contaminants to accumulate indoors. Indoor air pollutants include various combustion products from stoves, kerosene space heaters, and fireplaces, as well as volatile organic compounds (VOCs) from household products (e.g., paints, cleaning agents, and pesticides). Formaldehyde off-gassing from building products (especially particleboard and plywood) and from dry-cleaned textiles can accumulate in indoor air. Bacteria, viruses, molds, animal dander, dust mites, and pollen are biological contaminants that can cause disease and other health problems, especially if they build up in and are spread by central heating or cooling systems. Environmental tobacco smoke, also called secondhand smoke, is an indoor air pollutant in many homes, despite widespread knowledge about the harmful effects of smoking. Secondhand smoke contains many carcinogenic compounds as well as strong irritants. In some geographic regions, naturally occurring radon, a radioactive gas, can seep from the ground into buildings and accumulate to harmful levels. Exposure to all indoor air pollutants can be reduced by appropriate building construction and maintenance methods, limitations on pollutant sources, and provision of adequate ventilation.

Causes of Air Pollution

Burning of Fossil Fuels

Some of the most significant hidden costs of fossil fuels are from the air emissions that occur when they are burned. Unlike the extraction and transport stages, in which coal, oil, and natural gas can have very different types of impacts, all fossil fuels emit carbon dioxide and other harmful air pollutants when burned. These emissions lead to a wide variety of public health and environmental costs that are borne at the local, regional, national, and global levels.

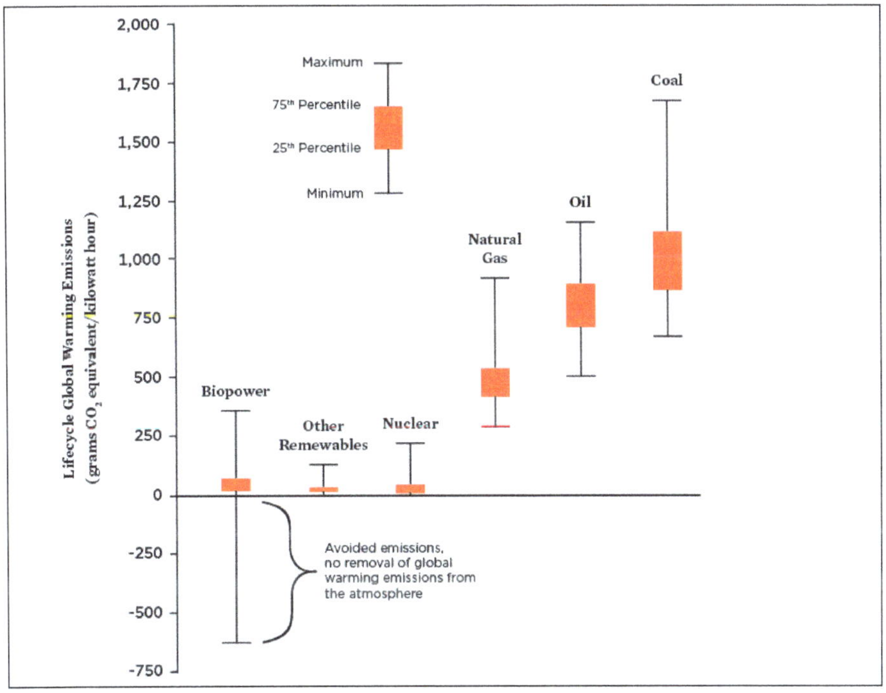

Comparing energy sources in terms of their global warming pollution.

The many environmental and public health risks associated with burning fossil fuels, the most serious in terms of its universal and potentially irreversible consequences is global warming. In 2014, approximately 78 percent of US global warming emissions were energy-related emissions of carbon dioxide. Of this, approximately 42 percent was from oil and other liquids, 32 percent from coal, and 27 percent from natural gas.

Non-fossil fuel energy generation technologies, like wind, solar, and geothermal, contributed less than 1 percent of the total energy related global warming emissions. Even when considering the full lifecycle carbon emissions of all energy sources, coal, oil, and natural gas clearly stand out with significantly higher greenhouse gas emissions.

The use of fossil fuels in transportation contributes almost 30 percent of all US global warming emissions, rivalling—and likely to surpass—the power sector. Burning fossil fuels emits a number of air pollutants that are harmful to both the environment and public health.

Sulfur dioxide (SO_2) emissions, primarily the result of burning coal, contribute to acid rain and the formation of harmful particulate matter. In addition, SO_2 emissions can exacerbate respiratory ailments, including asthma, nasal congestion, and pulmonary inflammation. In 2014, fossil fuel combustion at power plants accounted for 64 percent of US SO_2 emissions.

Nitrogen oxides (NOx) emissions, a byproduct of all fossil fuel combustion, contribute to acid rain and ground-level ozone (smog), which can burn lung tissue and can make people more susceptible to asthma, bronchitis, and other chronic respiratory diseases. Fossil fuel-powered transportation is the primary contributor to US NOx emissions.

Acid rain is formed when sulfur dioxide and nitrogen oxides mix with water, oxygen, and other chemicals in the atmosphere, leading to rain and other precipitation that is mildly acidic. Acidic precipitation increases the acidity of lakes and streams, which can be harmful to fish and other aquatic organisms. It can also damage trees and weaken forest ecosystems.

Particulate matter (soot) emissions produce haze and can cause chronic bronchitis, aggravated asthma, and elevated occurrence of premature death. In 2010, it is estimated that fine particle pollution from US coal plants resulted in 13,200 deaths, 9,700 hospitalizations, and 20,000 heart attacks. The impacts are particularly severe among the young, the elderly, and those who suffer from respiratory disease. The total health cost was estimated to be more than $100 billion per year.

Coal-fired power plants are the largest source of mercury emissions to the air in the United States. As airborne mercury settles onto the ground, it washes into bodies of water where it accumulates in fish, and subsequently passes through the food chain to birds and other animals. The consumption of mercury-laden fish by pregnant women has been associated with neurological and neurobehavioral effects in infants. Young children are also at risk.

Air pollution in Los Angeles.

A number of studies have sought to quantify the health costs associated with fossil fuel-related air pollution. The National Academy of Sciences assessed the costs of SO_2, NOx, and particulate matter air pollution from coal and reported an annual cost of $62 billion for 2005 —approximately 3.2 cents per kilowatt-hour (kWh). A separate study estimated that the pollution costs from coal combustion, including the effects of volatile organic compounds (VOCs) and ozone, was approximately $187 billion annually, or 9.3 cents per kWh.

A 2013 study also assessed the economic impacts of fossil fuel use, including illnesses, premature mortality, workdays lost, and direct costs to the healthcare system associated with emissions of particulates, NOx, and SO_2. This study found an average economic cost (or "public health added cost") of 32 cents per kWh for coal, 13 cents per kWh for oil, and 2 cents per kWh for natural gas. While cost estimates vary depending on each study's scope and assumptions, together they demonstrate the significant and real economic costs that unpriced air emissions impose on society.

Fossil fuel transportation emissions represent the largest single source of toxic air pollution in the U.S., accounting for over a third of carbon monoxide (CO) and NOx emissions.

Air Pollution Caused by Industries

Before we examine air pollution from industries, we need to understand the natural conditions and constituents of the atmosphere. The atmosphere, or air, is normally composed of about 79 percent nitrogen, 20 percent oxygen, and 1 percent mixture of carbon dioxide, water vapor, and small quantities of several gases. A list of these contents is given in table along with their average concentrations and approximate residence time. Based primarily on temperature, the atmosphere is divided into four layers: (I) Troposphere is the layer between the ground and a height of approximately 15 km at the equator and 10 km at the poles. In the troposphere, the temperature typically decreases at the rate of 6.5 °C/km, the air is relatively well mixed, and thermally driven convection is relatively strong. The troposphere is the layer that contains most of our air pollutants (II) Stratosphere is the layer between the troposphere and approximately 50 km. There is little bulk mixing in this layer. The temperature is relatively constant except in the upper stratosphere, where the absorption of UV radiation by ozone causes the temperature to rise to approximately 0 °C. (III) Mesosphere is the layer between 50 and 85 km within which the temperature decreases almost linearly to approximately 175 K. (IV) Thermosphere is the uppermost layer of the atmosphere. The temperature increases to approximately 1500 K, but it cannot be measured with a thermometer because the molecular concentrations are very small about 10^{13} molecules/cm^3 by comparison to the molecular concentration of 2.5×10^{19} molecules/cm^3 at the earth's surface.

Pollution there are several natural sources that degrade the quality of the air, such as gases and particles from volcanoes, dust from natural sources or odors from decomposition of dead plants and animals. However, human beings do not control these phenomena, and they do not fit our definition of pollution.

Main air Pollutants and Industrial Emissions Inventory

Major Air Pollutants

Air pollutants can be divided into two main groups—particulate and gaseous. The former group includes solid airborne particulates such as dust, fly ash, smoke, fog, soot, and fumes. Gaseous pollutants include carbon monoxide, hydrocarbon, oxides of sulfur and oxides of nitrogen. These pollutants are known as primary air pollutants. These materials may interact with one another in the presence of an energy source to form new secondary air pollutants such as ozone and other very reactive materials. Secondary air pollutants also form from reactions with natural chemicals in the atmosphere.

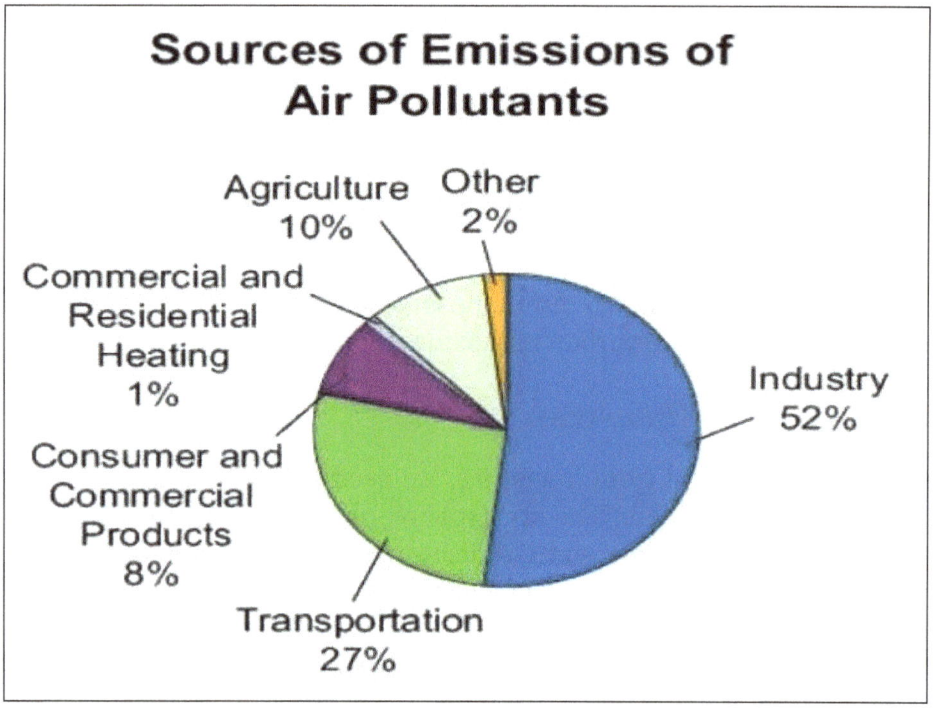

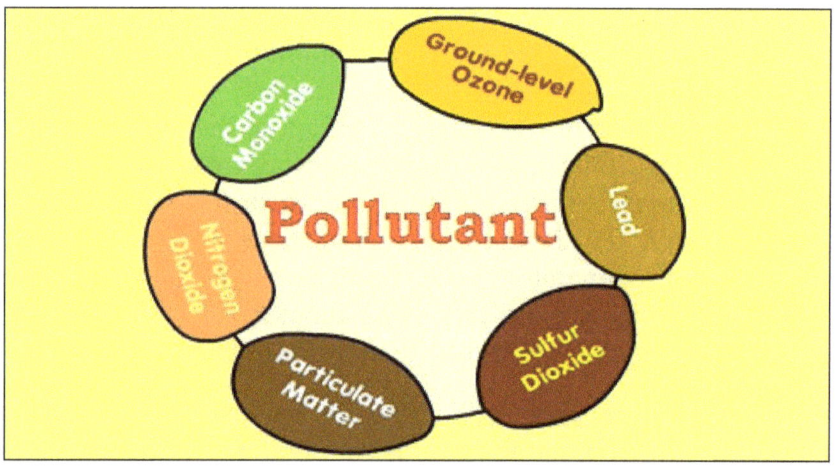

Table: Sources of primary air pollutants.

Pollutants	Sources
Carbon monoxide(CO)	Incomplete burning of fossil fuels Tobacco smoke.
Hydrogen Carbon	Incomplete burning of fossil fuels Tobacco smoke, Chemical processes.
Particulate	Burning fossil fuels, Construction operations Industrial waste.
Oxide of sulfur (SOx)	Burning fossil fuels, Chemical processes Smelting ores.
Oxide of nitrogen(NOx)	Burning fossil fuels, Oxide of N_2 in atmosphere.

Oxides of Sulfur (SOx): Oxides of sulfur (SOx) are compounds of sulfur and oxygen that are mostly produced when sulfur-containing fossil fuels are burned. Coal and oil were produced from organisms that had sulfur in their living structure. When the coal or oil was formed, some of the sulfur was incorporated into the fossil fuel. The sulfur is released as sulfur oxide when the fuel is burned. Simultaneously, a large quantity of sulfur oxides is released along with the exhaust gas in metallurgy and vitriol producing process. The reactions involved are as follows:

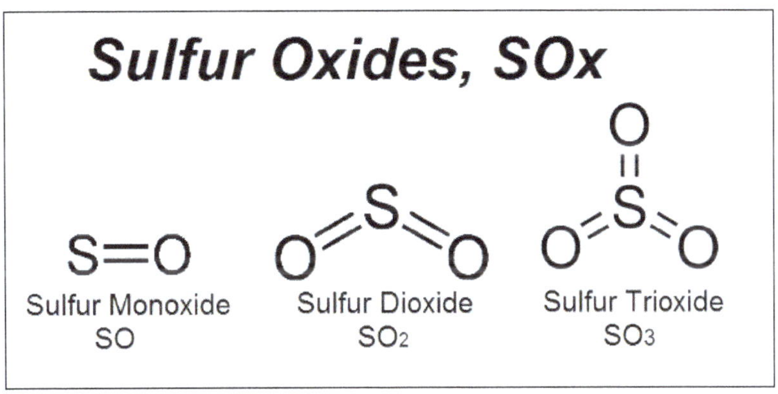

Burning of elemental sulfur: $S+O_2=SO_2$, $SO_2+0.5O2=SO_3$

Burning of iron sulfur ore: $4FeS_2+11O_2-2Fe_2O_3+8SO_2$, $SO_2+0.5O_2=SO_3$

Burning of organic sulfur compounds:

$$CH_3CH_2CH_2CH_2SH \rightarrow H_2S+2H_2+2C+C_2H_4$$

$$2H_2S+3O_2 \rightarrow 2SO_2+2H_2O$$

Oxides of nitrogen (NOx): As a category of primary air pollutants, oxides of nitrogen (NOx) emitted from industries are a mixture of nitrogen oxide (NO) and nitrogen dioxide (NO_2). About half of NOx, or up to 5.21×10^6 t/a, is produced by human activities such as fossil free burning, production of nitric acid and nitrogenous fertilizer, and other industrial processes.

There are two kinds of NOx from fuel combustion: one is thermal NOx, from N_2 and O_2 reaction when burned at high temperature, and the other is fuel NOx, which is the decomposed and oxidized compounds of nitrogenous compounds in fuel such as pyridine (C_5H_5N), piperidine ($C_5H_{11}N$) and carbazole ($C_{12}H_9N$) at high temperature. Most of fuel NOx is NO, less than 10% NO is oxidized to NO_2 in flue gas.

Carbon Monoxide (CO): Carbon monoxide is produced when organic materials, such as gasoline, coal, wood, and trash, are incompletely burned. The reactions are following:

$$C+0.5O_2 \rightarrow COC+CO_2 \rightarrow 2CO$$

Due to the lack of oxygen, the oxidation rate from CO to CO_2 is very slow. Improvements in combustion equipment and technology cause the discharge of CO from stationary sources to decrease gradually. Table below shows the contribution of CO from different sources.

Source	Discharge	Source	Discharge
Industry	300-500	Oxidization of methane	400-1000
Burning of biomass	300-700	Oxidization of other	
Activity of creature	60-160	Hydrocarbon	200-600
Discharge of sea	20-200	Total	1800-2700

Hydrocarbon (HC): Hydrocarbons are a large family of compounds, which include most organic compounds with less than about 12 carbon atoms. They are either evaporated from fuel or are remnants of fuel that did not burn completely. Table below shows the contribution of hydrocarbon from different sources.

Source	Discharge	Source	Discharge
Coal	2.9	Dissolvent	10
Thermal power	0.2	Burning of rubbish	25
Industrial	0.7	Burning of lumber	0.7
Resident and commerce	2.0	Forest fire	1.2
Petroleum	48.4	Total	88.2
Petroleum refinery	6.3		
Gasoline	34		
Diesel oil	0.1		
Heavy oil	0.2		
Evaporation of oil	7.8		

Particulates: Particulates, small pieces of solid materials dispersed into the atmosphere, constitute the third largest category of air pollutants. Smoke particles from fires, bits of asbestos from brake linings and insulation, dust particles, and ash from industrial plants contribute to the particulate load.

Industrial Emissions Inventory

Due to the wide variety of industrial processes, reliable determinations of industrial emissions have been found to depend on individual types of treatment. This is expensive and time-consuming, as it involves detailed study and testing of specific plants and processes. In some cases, data on industrial emissions could be obtained by means of questionnaires sent to the individual plants requesting information on processes, types and quantities of process exhaust, air cleaning equipment and fuel consumption. Frequently, the industries themselves do not have on-line data of this sort. Consequently, insufficient or incomplete results can be expected. Information on air emissions can also be obtained by using engineering estimates which are based on the raw materials or fuel consumed in a given process and reported figures on types and rate of pollutants emitted per unit weight of raw material or fuel used in a particular process. Unfortunately, there does not exist a sufficient body of readily available and reliable information on process emissions derived from actual tests, so it is difficult to obtain satisfactory results. The detailed procedure for making an industrial survey involves the use of a computer card identifying each establishment and its location. A detailed questionnaire sent to each establishment to collect information, which includes data on plant size, operating schedule, fuel usage, processes, and solvent usage. This procedure may provide more reliable results.

Emission factors, which have been applied to source data to yield emission data for SOx, NOx, HC, CO, particulates and so on, exist for a wide variety of industrial processes such as fuel burning, chemical production, manufacturing processes, and solvent usage.

Classification of air pollution sources and emissions are showed in table below.

Source type	Category	Examples	Pollutants
Dust production Process	Crushing, grinding screening, demolition milling	Road mix plants Urban renewal Grain elevators	Mineral and organic particulates
Combustion	Fuel burning	Home heating unit Power plants	SOx, NOx, CO, smoke, fly-ash organic vapors,
	Motor vehicles	Autos, buses and trucks Community and apartment	metal oxide particles and odors
	Refuse burning	house incinerator, open burning dums	
Manufacturing processes	Metallurgical plants	Smelters, steel mills, aluminum refineries	Metal fume, fluorides SOx
	Chemical plants	Petroleum refineries, pulp mills, super phosphate fertilizer plants, cement plants	H_2S, Sox, F, organic vapor, particle, odors
	Waste recovery	Metal scarp yards, auto body burning, rendering-plants	Smoke, soot, organic vapor,odors

Solvent	Spray painting	Furniture and appliance finishing	Organic phosphates, chlorinated
	Inks	Photogravure and printing	hydrocarbons, As, Pb
	Solvent cleaning	Dry cleaning, degreasing	

Combustion Sources and their Emission Characterization

Combustion refers to the rapid oxidation of substances (usually referred to as fuels) with the evolution of heat. Most fuels are a mixture of chemical compounds referred to as hydrocarbons, such as coal, fuel oil and natural gas, which, when burned, are converted into final products of carbon dioxide and water. The reaction products of the combustion process are collectively called combustion products, and include CO2, H2O, CO, HC, SO2, SO3 and NOx. In the case of fuels containing halides (usually chlorine and fluorine), the combustion products may contain hydrochloric or hydrofluoric acids (HCl, HF) and trace quantities of organohalides such as dioxins and furans.

In general, the combustion process can be represented by the following equation.

Fuel + Oxidant + Diluent → Combustion Products + heat $CxHy+aO_2+bN_2 \rightarrow xCO_2+yH_2O+zNOx+mSO_2+nN_2+ash$

Most combustion systems operate with an overall excess of air to ensure complete combustion. The excess air would then function as an additional diluent, contribute to the combustion products and result in the emission of gaseous or particulate pollutants.

Based on the physical states in which they are found in nature, fuels can be classified as solid, liquid, or gaseous, with coal, fuel oil, and natural gas being the three most common examples of each. The following sections describe these three fuels and their emission characteristics.

Coal Combustion

Coal is the most readily available energy source throughout the world. While coal is important to the development of industry, it results in air pollution when burned. During combustion, the coal is subject to either or both oxidizing and reducing conditions because of the presence of oxygen, carbon, and carbon monoxide in the bed. The air pollutants from the combustion of coal include particulates, SOx, NOx, HC, and CO. Amounts of each are highly dependent on the type of coal, combustion process, and operation of the process. The particulates consist mostly of ash particles. They also include unburned coal blown out during the turbulent combustion. Typical pulverized coal-fired boiler flue gas contains 2-10ppm CO, 3.9-5.1% O_2, and 14.3-15.3% CO_2. Gaseous emissions of sulfur oxides in the process of coal combustion are largely SO_2 and to a much smaller extent, SO_3 and gaseous sulfates. These compounds form as the organic and pyritic sulfur in the coal is oxidized during the combustion process. On average, 98% of the sulfur present in bituminous coal will be emitted as gaseous sulfur oxides, whereas somewhat less will be emitted when sub-bituminous coal is fired. Generally, boiler size, firing configuration, and boiler operations have little effect on the percent conversion of fuel sulfur to oxides of sulfur. Nitrogen oxides are produced in the combustion process by at least two different mechanisms: one source is from the molecular nitrogen in the combustion air (thermal NOx), the other is from the nitrogen in the fuel being burnt (fuel NOx). Generally, the amount of nitrogen in the fuel is relatively small compared with that in the combustion air, therefore, fuel NOx is only a small part of total NOx. The relative contributions of these two formation pathways de-pend on the combustion conditions, the type of boiler, and the type of fuel being burnt. Large coalfired boilers emit more NOx than any other combustion source. The combustion paradox is that the size increase of boilers will increase nitric oxide but decrease CO and HC. The reverse is true. Volatile organic compounds (VOCs) and carbon monoxide (CO) are unburned gaseous combustibles that are also emitted from coalfired boilers, but generally in quite small amounts. However, during start-up, temporary upsets, or other conditions preventing com-plete combustion, unburned combustible emissions may increase dramatically. The VOC and CO emissions per unit of fuel that is burned are normally lower from pulverized-coal or cyclone furnaces than from smaller stokers and hand-fired units where operating conditions are not as well controlled.

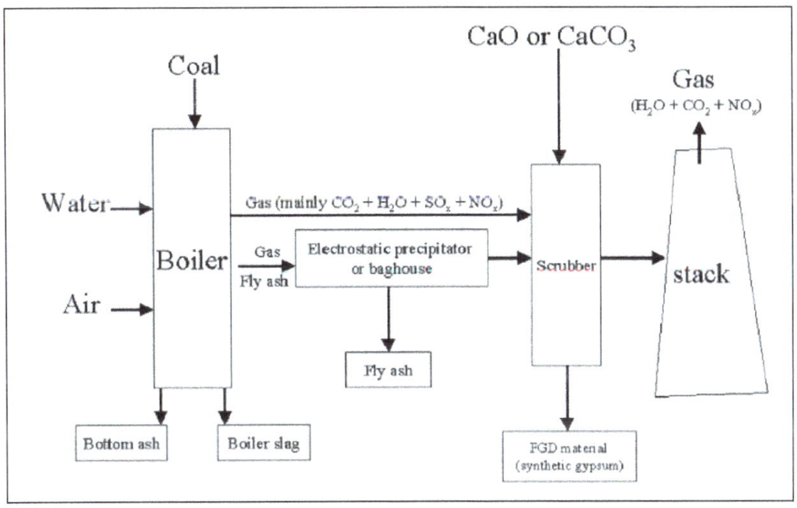

Oil Combustion

The most common liquid fuel used in utility processes is fuel oil derived from crude petroleum. Fuel oils are broadly classified into distillate kerosene and residual fuel oils. The former is mostly used for domestic and small commercial applications; the latter is used in utility and industrial boilers. Emissions from fuel oil combustion depend on the grade and composition of the oil, the type and size of the boiler, the firing practices used, and the level of equipment maintenance. Properly designed and operated oil-fired boilers produce less uncontrolled air pollution emission than coal-fired boilers but more than gas-fired boilers.

Table: Typical uncontrolled emission from oil combustion.

Boiler	Fuel oil	Uncontrolled Emissions, g/l of Fuel				
		particulate	SOx[a]	CO	HC[b]	NOx[c]
Utility or large						
Industrial	Residual	1	19.2S	0.4	0.25	0.12
Commercial or	Residual	2.75	19.2S	0.5	0.35	0.12
industrial	Distillate	1.8	17.2S	0.5	0.35	0.25
Small commercial						
or domestic	Distillate	1.2	17.2S	0.6	0.35	0.25

S = weight percent sulfur in oil; pollutant weight is reported as SO_2.

Pollutants weight is reported as CH_4

Pollutants weight is reported as NO_2; use one-half this value for tangentially fired boilers.

Horizontal firing and tangential firing are the two basic arrangements used in oil combustion systems. The oil is directed toward the center of the furnace in a horizontally fired system resulting in a hotter flame and the formation of about twice the NO_x as is formed by tangentially fired oil furnaces. Large boilers burn hotter and produce more NO_x than small one. About 0.65% of sulfur dioxide converts to sulfur trioxide in an oil-fired boiler.

Forest Fires

Forest fire release pollutants like smoke, sulphur dioxide, nitrogen dioxide, ozone, particulate matter and carbon monoxide into the atmosphere. The bigger the bushfire, the bigger emissions.

In small amounts, carbon monoxide is present human bodies. The problem is that when inhaled in large amounts, and large amount of this gas are emitted on bushfires, it becomes toxic and can cause death.

Particulates on forest fire are either solid or liquid and consist of soot, tars, and other volatile organic substances. These particles can have different sizes: PM10, PM2.5, PM0.3 and smaller. They can cause cardiovascular and respiratory problems if they penetrate deeply into humans lungs.

Nitrogen oxides and sulphur dioxides are minor problems when it comes to forest fire. Nitrogen oxides are only released in big bushfires, as they only appear at greater temperatures than 1,500 degree centigrade. On the other hand, except when peat and muck soil are involved, sulphur dioxide only appears to be less than 0,2%.

Health Effects of Wildfires

Wildfire in California during this summer 2018 is causing hazardous air conditions across the state, prompting air quality alerts and forcing many residents to take refuge indoors to avoid unhealthy exposure to bad air.

It is not just about foggy and hazy skies, human' risks and exposure are real. The air pollution from wildfires includes huge percentage of particles PM10, PM2.5, PM0.3 and thinner. Therefore, the health effects of wildfires are very similar to the ones explained during the blog post about Particulate Matter.

Firemen wildfire.

These health effects of forest fires include coughing, sore throats, extreme wheezing, cardiovascular illnesses and problems into lungs and bloodstream. Of course, sensible groups like children or elderly, are more likely to acquire these symptoms.

However, what do you have to do in the event of a forest fire? Firstly, you should be worried about the dangers of wildfires, so stay tuned to a local station for recommendations. Additionally, to protect yourself from air pollution: wear some kind of pollution mask, avoid staying outdoors and close all the opening from your house to prevent air pollution to enter. Finally, avoiding riding cars will help reduce SMOG, as this is produced with the mixing of combustion gases and the air pollution produced in forest fires.

Effects of Air Pollution

Global Warming

Global warming is the phenomenon of increasing average air temperatures near the surface of Earth over the past one to two centuries. Climate scientists have since the mid-20th century gathered detailed observations of various weather phenomena (such as temperatures, precipitation, and storms) and of related influences on climate (such as ocean currents and the atmosphere's chemical composition). These data indicate that Earth's climate has changed over almost every conceivable timescale since the beginning of geologic time and that the influence of human activities since at least the beginning of the Industrial Revolution has been deeply woven into the very fabric of climate change.

Giving voice to a growing conviction of most of the scientific community, the Intergovernmental Panel on Climate Change (IPCC) was formed in 1988 by the World Meteorological Organization (WMO) and the United Nations Environment Program (UNEP). In 2013 the IPCC reported that the interval between 1880 and 2012 saw an increase in global average surface temperature of approximately 0.9 °C (1.5 °F). The increase is closer to 1.1 °C (2.0 °F) when measured relative to the preindustrial mean temperature.

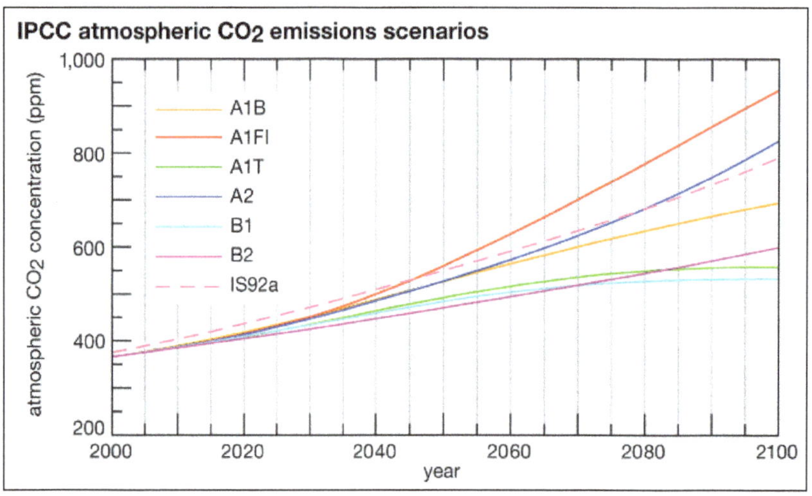

Global warming scenarios.

Graph of the predicted increase in Earth's average surface temperature according to a series of climate change scenarios that assume different levels of economic development, population growth, and fossil fuel use. The assumptions made by each scenario are given at the bottom of the graph.

A special report produced by the IPCC in 2018 honed this estimate further, noting that human beings and human activities have been responsible for a worldwide average temperature increase of between 0.8 and 1.2 °C (1.4 and 2.2 °F) of global warming since preindustrial times, and most of the warming observed over the second half of the 20th century could be attributed to human activities. It predicted that the global mean surface temperature would increase between 3 and 4 °C (5.4 and 7.2 °F) by 2100 relative to the 1986–2005 average should carbon emissions continue at their current rate. The predicted rise in temperature was based on a range of possible scenarios that accounted for future greenhouse gas emissions and mitigation (severity reduction) measures and on uncertainties in the model projections. Some of the main uncertainties include the precise role of feedback processes and the impacts of industrial pollutants known as aerosols, which may offset some warming.

Many climate scientists agree that significant societal, economic, and ecological damage would result if global average temperatures rose by more than 2 °C (3.6 °F) in such a short time. Such damage would include increased extinction of many plant and animal species, shifts in patterns of agriculture, and rising sea levels. By 2015 all but a few national governments had begun the process of instituting carbon reduction plans as part of the Paris Agreement, a treaty designed to help countries keep global warming to 1.5 °C (2.7 °F) above preindustrial levels in order to avoid the worst of the predicted effects. Authors of a special report published by the IPCC in 2018 noted that should carbon emissions continue at their present rate, the increase in average near-surface air temperatures would reach 1.5 °C sometime between 2030 and 2052. Past IPCC assessments reported that the global average sea level rose by some 19–21 cm (7.5–8.3 inches) between 1901 and 2010 and that sea levels rose faster in the second half of the 20th century than in the first half. It also predicted, again depending on a wide range of scenarios, that the global average sea level would rise 26–77 cm (10.2–30.3 inches) relative to the 1986–2005 average by 2100 for global warming of 1.5 °C, an average of 10 cm (3.9 inches) less than what would be expected if warming rose to 2 °C (3.6 °F) above preindustrial levels.

The scenarios referred to above depend mainly on future concentrations of certain trace gases, called greenhouse gases, that have been injected into the lower atmosphere in increasing amounts through the burning of fossil fuels for industry, transportation, and residential uses. Modern global warming is the result of an increase in magnitude of the so-called greenhouse effect, a warming of Earth's surface and lower atmosphere caused by the presence of water vapour, carbon dioxide, methane, nitrous oxides, and other greenhouse gases. In 2014 the IPCC reported that concentrations of carbon dioxide, methane, and nitrous oxides in the atmosphere surpassed those found in ice cores dating back 800,000 years.

All these gases, carbon dioxide is the most important, both for its role in the greenhouse effect and for its role in the human economy. It has been estimated that, at the beginning of the industrial age in the mid-18th century, carbon dioxide concentrations in the atmosphere were roughly 280 parts per million (ppm). By the middle of 2018 they had risen to 406 ppm, and, if fossil fuels continue to be burned at current rates, they are projected to reach 550 ppm by the mid-21st century—essentially, a doubling of carbon dioxide concentrations in 300 years.

A vigorous debate is in progress over the extent and seriousness of rising surface temperatures, the effects of past and future warming on human life, and the need for action to reduce future warming and deal with its consequences. This article provides an overview of the scientific background and public policy debate related to the subject of global warming. It considers the causes of rising near-surface air temperatures, the influencing factors, the process of climate research and forecasting, the possible ecological and social impacts of rising temperatures, and the public policy developments since the mid-20th century.

Climatic Variation since the Last Glaciation

Global warming is related to the more general phenomenon of climate change, which refers to changes in the totality of attributes that define climate. In addition to changes in air temperature, climate change involves changes to precipitation patterns, winds, ocean currents, and other measures of Earth's climate. Normally, climate change can be viewed as the combination of various natural forces occurring over diverse timescales. Since the advent of human civilization, climate change has involved an "anthropogenic," or exclusively human-caused, element, and this anthropogenic element has become more important in the industrial period of the past two centuries. The term global warming is used specifically to refer to any warming of near-surface air during the past two centuries that can be traced to anthropogenic causes.

A series of photographs of the Grinnell Glacier taken from the summit of Mount Gould in Glacier National Park, Montana, in 1938, 1981, 1998, and 2006 (from left to right). In 1938 the Grinnell Glacier filled the entire area at the bottom of the image. By 2006 it had largely disappeared from this view.

To define the concepts of global warming and climate change properly, it is first necessary to recognize that the climate of Earth has varied across many timescales, ranging from an individual human life span to billions of years. This variable climate history is typically classified in terms of "regimes" or "epochs." For instance, the Pleistocene glacial epoch (about 2,600,000 to 11,700 years ago) was marked by substantial variations in the global extent of glaciers and ice sheets. These variations took place on timescales of tens to hundreds of millennia and were driven by changes in the distribution of solar radiation across Earth's surface. The distribution of solar radiation is known as the insolation pattern, and it is strongly affected by the geometry of Earth's orbit around the Sun and by the orientation, or tilt, of Earth's axis relative to the direct rays of the Sun.

Worldwide, the most recent glacial period, or ice age, culminated about 21,000 years ago in what is often called the Last Glacial Maximum. During this time, continental ice sheets extended well into the middle latitude regions of Europe and North America, reaching as far south as present-day London and New York City. Global annual mean temperature appears to have been about 4–5 °C (7–9 °F) colder than in the mid-20th century. It is important to remember that these figures are a global average. In fact, during the height of this last ice age, Earth's climate was characterized by greater cooling at higher latitudes (that is, toward the poles) and relatively little cooling over large parts of the tropical oceans (near the Equator). This glacial interval terminated abruptly about 11,700 years ago and was followed by the subsequent relatively ice-free period known as the Holocene Epoch. The modern period of Earth's history is conventionally defined as residing within the Holocene. However, some scientists have argued that the Holocene Epoch terminated in the relatively recent past and that Earth currently resides in a climatic interval that could justly be called the Anthropocene Epoch—that is, a period during which humans have exerted a dominant influence over climate.

Though less dramatic than the climate changes that occurred during the Pleistocene Epoch, significant variations in global climate have nonetheless taken place over the course of the Holocene. During the early Holocene, roughly 9,000 years ago, atmospheric circulation and precipitation patterns appear to have been substantially different from those of today. For example, there is evidence for relatively wet conditions in what is now the Sahara Desert. The change from one climatic regime to another was caused by only modest changes in the pattern of insolation within the Holocene interval as well as the interaction of these patterns with large-scale climate phenomena such as monsoons and El Niño/Southern Oscillation (ENSO).

During the middle Holocene, some 5,000–7,000 years ago, conditions appear to have been relatively warm—indeed, perhaps warmer than today in some parts of the world and during certain seasons. For this reason, this interval is sometimes referred to as the Mid-Holocene Climatic Optimum. The relative warmth of average near-surface air temperatures at this time, however, is somewhat unclear. Changes in the pattern of insolation favoured warmer summers at higher latitudes in the Northern Hemisphere, but these changes also produced cooler winters in the Northern Hemisphere and relatively cool conditions year-round in the tropics. Any overall hemispheric or global mean temperature changes thus reflected a balance between competing seasonal and regional changes. In fact, recent theoretical climate model studies suggest that global mean temperatures during the middle Holocene were probably 0.2–0.3 °C (0.4–0.5 °F) colder than average late 20th-century conditions.

Over subsequent millennia, conditions appear to have cooled relative to middle Holocene levels. This period has sometimes been referred to as the "Neoglacial." In the middle latitudes this cooling trend was associated with intermittent periods of advancing and retreating mountain glaciers reminiscent of (though far more modest than) the more substantial advance and retreat of the major continental ice sheets of the Pleistocene climate epoch.

Causes of Global Warming

Greenhouse Effect

The average surface temperature of Earth is maintained by a balance of various forms of solar and terrestrial radiation. Solar radiation is often called "shortwave" radiation because the frequencies of the radiation are relatively high and the wavelengths relatively short—close to the visible portion of the electromagnetic spectrum. Terrestrial radiation, on the other hand, is often called "longwave" radiation because the frequencies are relatively low and the wavelengths relatively long—somewhere in the infrared part of the spectrum. Downward-moving solar energy is typically measured in watts per square metre. The energy of the total incoming solar radiation at the top of Earth's atmosphere (the so-called "solar constant") amounts roughly to 1,366 watts per square metre annually. Adjusting for the fact that only one-half of the planet's surface receives solar radiation at any given time, the average surface insolation is 342 watts per square metre annually.

The amount of solar radiation absorbed by Earth's surface is only a small fraction of the total solar radiation entering the atmosphere. For every 100 units of incoming solar radiation, roughly 30 units are reflected back to space by either clouds, the atmosphere, or reflective regions of Earth's surface. This reflective capacity is referred to as Earth's planetary albedo, and it need not remain fixed over time, since the spatial extent and distribution of reflective formations, such as clouds and ice cover, can change. The 70 units of solar radiation that are not reflected may be absorbed by the atmosphere, clouds, or the surface. In the absence of further complications, in order to maintain thermodynamic equilibrium, Earth's surface and atmosphere must radiate these same 70 units back to space. Earth's surface temperature (and that of the lower layer of the atmosphere essentially in contact with the surface) is tied to the magnitude of this emission of outgoing radiation according to the Stefan-Boltzmann law.

Earth's energy budget is further complicated by the greenhouse effect. Trace gases with certain chemical properties—the so-called greenhouse gases, mainly carbon dioxide (CO_2), methane (CH_4), and nitrous oxide (N_2O)—absorb some of the infrared radiation produced by Earth's surface. Because of this absorption, some fraction of the original 70 units does not directly escape to space. Because greenhouse gases emit the same amount of radiation they absorb and because this radiation is emitted equally in all directions (that is, as much downward as upward), the net effect of absorption by greenhouse gases is to increase the total amount of radiation emitted downward toward Earth's surface and lower atmosphere. To maintain equilibrium, Earth's surface and lower atmosphere must emit more radiation than the original 70 units. Consequently, the surface temperature must be higher. This process is not quite the same as that which governs a true greenhouse, but the end effect is similar. The presence of greenhouse gases in the atmosphere leads to a warming of the surface and lower part of the atmosphere (and a cooling higher up in the atmosphere) relative to what would be expected in the absence of greenhouse gases.

It is essential to distinguish the natural or background, greenhouse effect from the "enhanced" greenhouse effect associated with human activity. The natural greenhouse effect is associated with surface warming properties of natural constituents of Earth's atmosphere, especially water vapour, carbon dioxide, and methane. The existence of this effect is accepted by all scientists. Indeed, in its absence, Earth's average temperature would be approximately 33 °C (59 °F) colder than today, and Earth would be a frozen and likely uninhabitable planet. What has been subject to controversy is the so-called enhanced greenhouse effect, which is associated with increased concentrations of greenhouse gases caused by human activity. In particular, the burning of fossil fuels raises the concentrations of the major greenhouse gases in the atmosphere, and these higher concentrations have the potential to warm the atmosphere by several degrees.

Radiative Forcing

In light of the discussion above of the greenhouse effect, it is apparent that the temperature of Earth's surface and lower atmosphere may be modified in three ways: (1) through a net increase in the solar radiation entering at the top of Earth's atmosphere, (2) through a change in the fraction of the radiation reaching the surface, and (3) through a change in the concentration of greenhouse gases in the atmosphere. In each case the changes can be thought of in terms of "radiative forcing." As defined by the IPCC, radiative forcing is a measure of the influence a given climatic factor has on the amount of downward-directed radiant energy impinging upon Earth's surface. Climatic factors are divided between those caused primarily by human activity (such as greenhouse gas emissions and aerosol emissions) and those caused by natural forces (such as solar irradiance); then, for each factor, so-called forcing values are calculated for the time period between 1750 and the present day. "Positive forcing" is exerted by climatic factors that contribute to the warming of Earth's surface, whereas "negative forcing" is exerted by factors that cool Earth's surface.

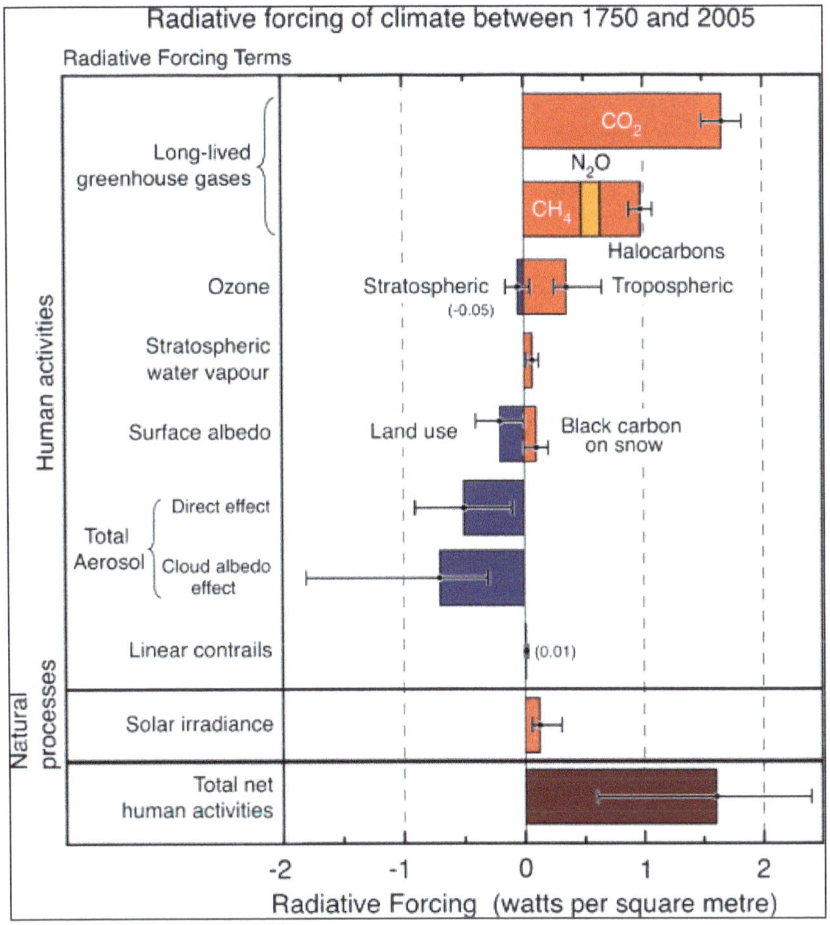

On average, about 342 watts of solar radiation strike each square metre of Earth's surface per year, and this quantity can in turn be related to a rise or fall in Earth's surface temperature. Temperatures at the surface may also rise or fall through a change in the distribution of terrestrial radiation (that is, radiation emitted by Earth) within the atmosphere. In some cases, radiative forcing has a natural origin, such as during explosive eruptions from volcanoes where vented gases and ash block some portion of solar radiation from the surface. In other cases, radiative forcing has an anthropogenic, or exclusively human, origin. For example, anthropogenic increases in carbon dioxide, methane, and nitrous oxide are estimated to account for 2.3 watts per square metre of positive radiative forcing. When all values of positive and negative radiative forcing are taken together and all interactions between climatic factors are accounted for, the total net increase in surface radiation due to human activities since the beginning of the Industrial Revolution is 1.6 watts per square metre.

Influences of Human Activity on Climate

Human activity has influenced global surface temperatures by changing the radiative balance governing the Earth on various timescales and at varying spatial scales. The most profound and well-known anthropogenic influence is the elevation of concentrations of greenhouse gases in the atmosphere. Humans also influence climate by changing the concentrations of aerosols and ozone and by modifying the land cover of Earth's surface.

Greenhouse Gases

Greenhouse gases warm Earth's surface by increasing the net downward longwave radiation reaching the surface. The relationship between atmospheric concentration of greenhouse gases and the associated positive radiative forcing of the surface is different for each gas. A complicated relationship exists between the chemical properties of each greenhouse gas and the relative amount of longwave radiation that each can absorb. What follows is a discussion of the radiative behaviour of each major greenhouse gas.

Factories that burn fossil fuels help to cause global warming.

Water Vapour

Water vapour is the most potent of the greenhouse gases in Earth's atmosphere, but its behaviour is fundamentally different from that of the other greenhouse gases. The primary role of water vapour is not as a direct agent of radiative forcing but rather as a climate feedback—that is, as a response within the climate system that influences the system's continued activity. This distinction arises from the fact that the amount of water vapour in the atmosphere cannot, in general, be directly modified by human behaviour but is instead set by air temperatures. The warmer the surface, the greater the evaporation rate of water from the surface. As a result, increased evaporation leads to a greater concentration of water vapour in the lower atmosphere capable of absorbing longwave radiation and emitting it downward.

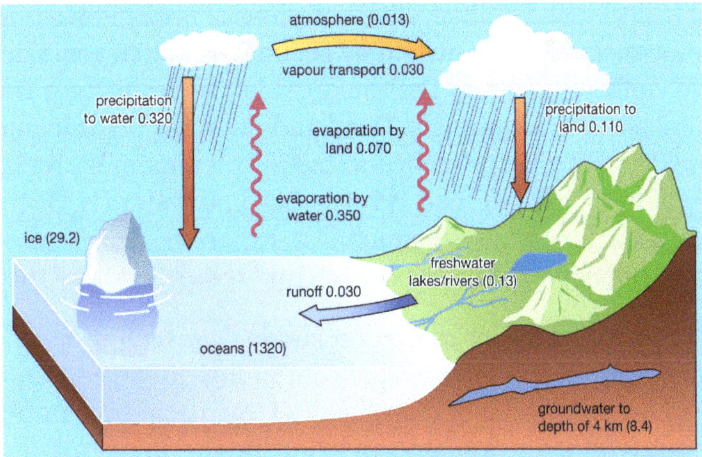

In the above figure, the present-day surface hydrologic cycle, in which water is transferred from the oceans through the atmosphere to the continents and back to the oceans over and beneath the land surface. The values in parentheses following the various forms of water (e.g., ice) refer to volumes in millions of cubic kilometres; those following the processes (e.g., precipitation) refer to their fluxes in millions of cubic kilometres of water per year.

Carbon Dioxide

Of the greenhouse gases, carbon dioxide (CO_2) is the most significant. Natural sources of atmospheric CO_2 include outgassing from volcanoes, the combustion and natural decay of organic matter, and respiration by aerobic (oxygen-using) organisms. These sources are balanced, on average, by a set of physical, chemical, or biological processes, called "sinks," that tend to remove CO_2 from the atmosphere. Significant natural sinks include terrestrial vegetation, which takes up CO_2 during the process of photosynthesis.

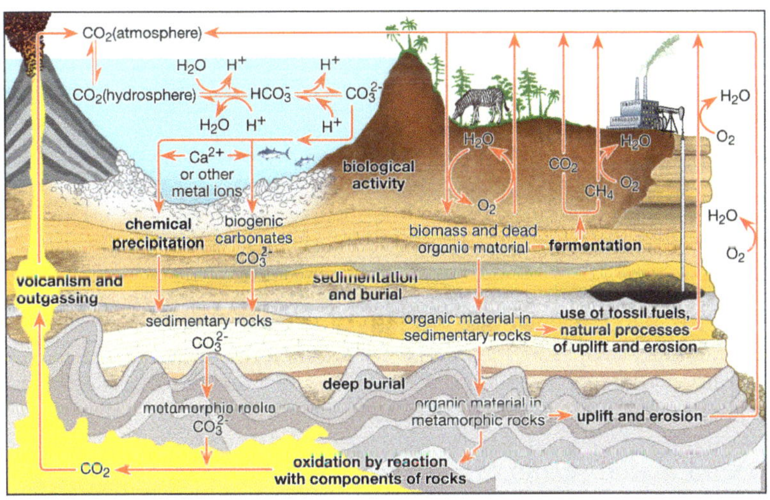

The carbon cycle Carbon is transported in various forms through the atmosphere, the hydrosphere, and geologic formations. One of the primary pathways for the exchange of carbon dioxide (CO_2) takes place between the atmosphere and the oceans; there a fraction of the CO_2 combines with water, forming carbonic acid (H_2CO_3) that subsequently loses hydrogen ions (H^+) to form bicarbonate

(HCO_3^-) and carbonate (CO_3^{2-}) ions. Mollusk shells or mineral precipitates that form by the reaction of calcium or other metal ions with carbonate may become buried in geologic strata and eventually release CO_2 through volcanic outgassing. Carbon dioxide also exchanges through photosynthesis in plants and through respiration in animals. Dead and decaying organic matter may ferment and release CO_2 or methane (CH_4) or may be incorporated into sedimentary rock, where it is converted to fossil fuels. Burning of hydrocarbon fuels returns CO_2 and water (H_2O) to the atmosphere. The biological and anthropogenic pathways are much faster than the geochemical pathways and, consequently, have a greater impact on the composition and temperature of the atmosphere.

A number of oceanic processes also act as carbon sinks. One such process, called the "solubility pump," involves the descent of surface seawater containing dissolved CO_2. Another process, the "biological pump," involves the uptake of dissolved CO_2 by marine vegetation and phytoplankton (small free-floating photosynthetic organisms) living in the upper ocean or by other marine organisms that use CO_2 to build skeletons and other structures made of calcium carbonate ($CaCO_3$). As these organisms expire and fall to the ocean floor, the carbon they contain is transported downward and eventually buried at depth. A long-term balance between these natural sources and sinks leads to the background, or natural, level of CO_2 in the atmosphere.

In contrast, human activities increase atmospheric CO_2 levels primarily through the burning of fossil fuels—principally oil and coal and secondarily natural gas, for use in transportation, heating, and the generation of electrical power—and through the production of cement. Other anthropogenic sources include the burning of forests and the clearing of land. Anthropogenic emissions currently account for the annual release of about 7 gigatons (7 billion tons) of carbon into the atmosphere. Anthropogenic emissions are equal to approximately 3 percent of the total emissions of CO_2 by natural sources, and this amplified carbon load from human activities far exceeds the offsetting capacity of natural sinks (by perhaps as much as 2–3 gigatons per year).

Deforestation Smoldering remains of a plot of deforested land in the Amazon Rainforest of Brazil. Annually, it is estimated that net global deforestation accounts for about two gigatons of carbon emissions to the atmosphere.

CO_2 consequently accumulated in the atmosphere at an average rate of 1.4 ppm per year between 1959 and 2006 and roughly 2.0 ppm per year between 2006 and 2018. Overall, this rate of accumulation has been linear (that is, uniform over time). However, certain current sinks, such as the oceans, could become sources in the future. This may lead to a situation in which the concentration of atmospheric CO_2 builds at an exponential rate (that is, its rate of increase is also increasing).

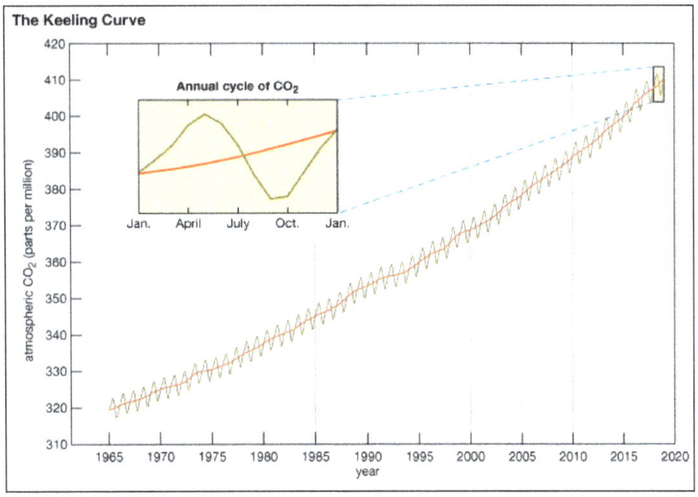

The Keeling Curve, named after American climate scientist Charles David Keeling, tracks changes in the concentration of carbon dioxide (CO_2) in Earth's atmosphere at a research station on Mauna Loa in Hawaii. Although these concentrations experience small seasonal fluctuations, the overall trend shows that CO_2 is increasing in the atmosphere.

The natural background level of carbon dioxide varies on timescales of millions of years because of slow changes in outgassing through volcanic activity. For example, roughly 100 million years ago, during the Cretaceous Period (145 million to 66 million years ago), CO_2 concentrations appear to have been several times higher than they are today (perhaps close to 2,000 ppm). Over the past 700,000 years, CO_2 concentrations have varied over a far smaller range (between roughly 180 and 300 ppm) in association with the same Earth orbital effects linked to the coming and going of the Pleistocene ice ages. By the early 21st century, CO_2 levels had reached 384 ppm, which is approximately 37 percent above the natural background level of roughly 280 ppm that existed at the beginning of the Industrial Revolution. Atmospheric CO_2 levels continued to increase, and by 2018 they had reached 410 ppm. Such levels are believed to be the highest in at least 800,000 years according to ice core measurements and may be the highest in at least 5 million years according to other lines of evidence.

Radiative forcing caused by carbon dioxide varies in an approximately logarithmic fashion with the concentration of that gas in the atmosphere. The logarithmic relationship occurs as the result of a saturation effect wherein it becomes increasingly difficult, as CO_2 concentrations increase, for additional CO_2 molecules to further influence the "infrared window" (a certain narrow band of wavelengths in the infrared region that is not absorbed by atmospheric gases). The logarithmic relationship predicts that the surface warming potential will rise by roughly the same amount for each doubling of CO_2 concentration. At current rates of fossil fuel use, a doubling of CO_2 concentrations over preindustrial levels is expected to take place by the middle of the 21st century (when CO_2 concentrations are projected to reach 560 ppm). A doubling of CO_2 concentrations would represent an increase of roughly 4 watts per square metre of radiative forcing. Given typical estimates of "climate sensitivity" in the absence of any offsetting factors, this energy increase would lead to a warming of 2 to 5 °C (3.6 to 9 °F) over preindustrial times. The total radiative forcing by anthropogenic CO_2 emissions since the beginning of the industrial age is approximately 1.66 watts per square metre.

Methane

Methane (CH_4) is the second most important greenhouse gas. CH_4 is more potent than CO_2 because the radiative forcing produced per molecule is greater. In addition, the infrared window is less saturated in the range of wavelengths of radiation absorbed by CH_4, so more molecules may fill in the region. However, CH_4 exists in far lower concentrations than CO_2 in the atmosphere, and its concentrations by volume in the atmosphere are generally measured in parts per billion (ppb) rather than ppm. CH_4 also has a considerably shorter residence time in the atmosphere than CO_2 (the residence time for CH_4 is roughly 10 years, compared with hundreds of years for CO_2).

Natural sources of methane include tropical and northern wetlands, methane-oxidizing bacteria that feed on organic material consumed by termites, volcanoes, seepage vents of the seafloor in regions rich with organic sediment, and methane hydrates trapped along the continental shelves of the oceans and in polar permafrost. The primary natural sink for methane is the atmosphere itself, as methane reacts readily with the hydroxyl radical (·OH) within the troposphere to form CO_2 and water vapour (H_2O). When CH_4 reaches the stratosphere, it is destroyed. Another natural sink is soil, where methane is oxidized by bacteria.

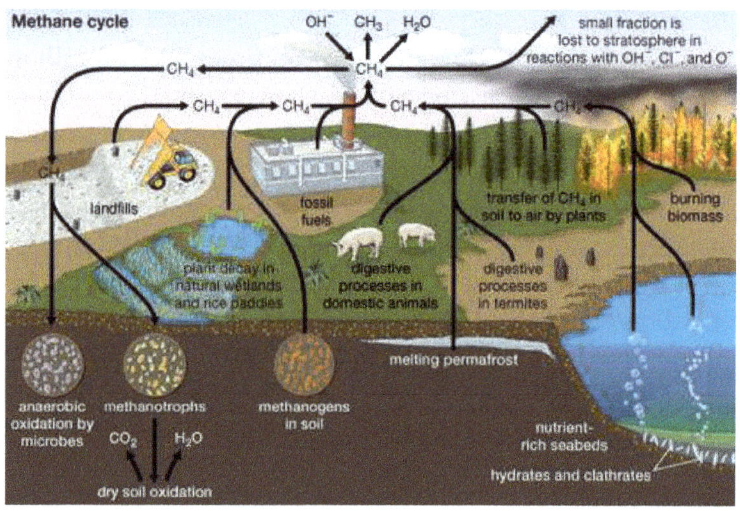

Methane cycle.

As with CO_2, human activity is increasing the CH_4 concentration faster than it can be offset by natural sinks. Anthropogenic sources currently account for approximately 70 percent of total annual emissions, leading to substantial increases in concentration over time. The major anthropogenic sources of atmospheric CH_4 are rice cultivation, livestock farming, the burning of coal and natural gas, the combustion of biomass, and the decomposition of organic matter in landfills. Future trends are particularly difficult to anticipate. This is in part due to an incomplete understanding of the climate feedbacks associated with CH_4 emissions. In addition it is difficult to predict how, as human populations grow, possible changes in livestock raising, rice cultivation, and energy utilization will influence CH_4 emissions.

It is believed that a sudden increase in the concentration of methane in the atmosphere was responsible for a warming event that raised average global temperatures by 4–8 °C (7.2–14.4 °F) over a few thousand years during the so-called Paleocene-Eocene Thermal Maximum, or PETM.

This episode took place roughly 55 million years ago, and the rise in CH_4 appears to have been related to a massive volcanic eruption that interacted with methane-containing flood deposits. As a result, large amounts of gaseous CH_4 were injected into the atmosphere. It is difficult to know precisely how high these concentrations were or how long they persisted. At very high concentrations, residence times of CH_4 in the atmosphere can become much greater than the nominal 10-year residence time that applies today. Nevertheless, it is likely that these concentrations reached several ppm during the PETM.

Methane concentrations have also varied over a smaller range (between roughly 350 and 800 ppb) in association with the Pleistocene ice age cycles. Preindustrial levels of CH_4 in the atmosphere were approximately 700 ppb, whereas levels exceeded 1,867 ppb in late 2018. (These concentrations are well above the natural levels observed for at least the past 650,000 years.) The net radiative forcing by anthropogenic CH_4 emissions is approximately 0.5 watt per square metre—or roughly one-third the radiative forcing of CO_2.

Surface-level Ozone and other Compounds

The next most significant greenhouse gas is surface, or low-level, ozone (O_3). Surface O_3 is a result of air pollution; it must be distinguished from naturally occurring stratospheric O_3, which has a very different role in the planetary radiation balance. The primary natural source of surface O_3 is the subsidence of stratospheric O_3 from the upper atmosphere. In contrast, the primary anthropogenic source of surface O_3 is photochemical reactions involving the atmospheric pollutant carbon monoxide (CO). The best estimates of the natural concentration of surface O_3 are 10 ppb, and the net radiative forcing due to anthropogenic emissions of surface O_3 is approximately 0.35 watt per square metre. Ozone concentrations can rise above unhealthy levels (that is, conditions where concentrations meet or exceed 70 ppb for eight hours or longer) in cities prone to photochemical smog.

Nitrous Oxides and Fluorinated Gases

Additional trace gases produced by industrial activity that have greenhouse properties include nitrous oxide (N_2O) and fluorinated gases (halocarbons), the latter including sulfur hexafluoride, hydrofluorocarbons (HFCs), and perfluorocarbons (PFCs). Nitrous oxide is responsible for 0.16 watt per square metre radiative forcing, while fluorinated gases are collectively responsible for 0.34 watt per square metre. Nitrous oxides have small background concentrations due to natural biological reactions in soil and water, whereas the fluorinated gases owe their existence almost entirely to industrial sources.

Aerosols

The production of aerosols represents an important anthropogenic radiative forcing of climate. Collectively, aerosols block—that is, reflect and absorb—a portion of incoming solar radiation, and this creates a negative radiative forcing. Aerosols are second only to greenhouse gases in relative importance in their impact on near-surface air temperatures. Unlike the decade-long residence times of the "well-mixed" greenhouse gases, such as CO_2 and CH_4, aerosols are readily flushed out of the atmosphere within days, either by rain or snow (wet deposition) or by settling out of the air (dry deposition). They must therefore be continually generated in order to produce a steady effect on radiative forcing.

Aerosols have the ability to influence climate directly by absorbing or reflecting incoming solar radiation, but they can also produce indirect effects on climate by modifying cloud formation or cloud properties. Most aerosols serve as condensation nuclei (surfaces upon which water vapour can condense to form clouds); however, darker-coloured aerosols may hinder cloud formation by absorbing sunlight and heating up the surrounding air. Aerosols can be transported thousands of kilometres from their sources of origin by winds and upper-level circulation in the atmosphere.

Perhaps the most important type of anthropogenic aerosol in radiative forcing is sulfate aerosol. It is produced from sulfur dioxide (SO_2) emissions associated with the burning of coal and oil. Since the late 1980s, global emissions of SO_2 have decreased from about 151.5 million tonnes (167.0 million tons) to less than 100 million tonnes (110.2 million tons) of sulfur per year.

Nitrate aerosol is not as important as sulfate aerosol, but it has the potential to become a significant source of negative forcing. One major source of nitrate aerosol is smog (the combination of ozone with oxides of nitrogen in the lower atmosphere) released from the incomplete burning of fuel in internal-combustion engines. Another source is ammonia (NH_3), which is often used in fertilizers or released by the burning of plants and other organic materials. If greater amounts of atmospheric nitrogen are converted to ammonia and agricultural ammonia emissions continue to increase as projected, the influence of nitrate aerosols on radiative forcing is expected to grow.

Both sulfate and nitrate aerosols act primarily by reflecting incoming solar radiation, thereby reducing the amount of sunlight reaching the surface. Most aerosols, unlike greenhouse gases, impart a cooling rather than warming influence on Earth's surface. One prominent exception is carbonaceous aerosols such as carbon black or soot, which are produced by the burning of fossil fuels and biomass. Carbon black tends to absorb rather than reflect incident solar radiation, and so it has a warming impact on the lower atmosphere, where it resides. Because of its absorptive properties, carbon black is also capable of having an additional indirect effect on climate. Through its deposition in snowfall, it can decrease the albedo of snow cover. This reduction in the amount of solar radiation reflected back to space by snow surfaces creates a minor positive radiative forcing.

Natural forms of aerosol include windblown mineral dust generated in arid and semiarid regions and sea salt produced by the action of waves breaking in the ocean. Changes to wind patterns as a result of climate modification could alter the emissions of these aerosols. The influence of climate change on regional patterns of aridity could shift both the sources and the destinations of dust clouds. In addition, since the concentration of sea salt aerosol, or sea aerosol, increases with the strength of the winds near the ocean surface, changes in wind speed due to global warming and climate change could influence the concentration of sea salt aerosol. For example, some studies suggest that climate change might lead to stronger winds over parts of the North Atlantic Ocean. Areas with stronger winds may experience an increase in the concentration of sea salt aerosol.

Other natural sources of aerosols include volcanic eruptions, which produce sulfate aerosol, and biogenic sources (e.g., phytoplankton), which produce dimethyl sulfide (DMS). Other important biogenic aerosols, such as terpenes, are produced naturally by certain kinds of trees or other plants. For example, the dense forests of the Blue Ridge Mountains of Virginia in the United States emit terpenes during the summer months, which in turn interact with the high humidity and warm temperatures to produce a natural photochemical smog. Anthropogenic pollutants such as nitrate and ozone, both of which serve as precursor molecules for the generation of biogenic aerosol, appear to have increased the rate of production of these aerosols severalfold. This process appears to be responsible for some of the increased aerosol pollution in regions undergoing rapid urbanization.

Human activity has greatly increased the amount of aerosol in the atmosphere compared with the background levels of preindustrial times. In contrast to the global effects of greenhouse gases, the impact of anthropogenic aerosols is confined primarily to the Northern Hemisphere, where most of the world's industrial activity occurs. The pattern of increases in anthropogenic aerosol over time is also somewhat different from that of greenhouse gases. During the middle of the 20th century, there was a substantial increase in aerosol emissions. This appears to have been at least partially responsible for a cessation of surface warming that took place in the Northern Hemisphere from the 1940s through the 1970s. Since that time, aerosol emissions have leveled off due to antipollution measures undertaken in the industrialized countries since the 1960s. Aerosol emissions may rise in the future, however, as a result of the rapid emergence of coal-fired electric power generation in China and India.

The total radiative forcing of all anthropogenic aerosols is approximately −1.2 watts per square metre. Of this total, −0.5 watt per square metre comes from direct effects (such as the reflection of solar energy back into space), and −0.7 watt per square metre comes from indirect effects (such as the influence of aerosols on cloud formation). This negative radiative forcing represents an offset of roughly 40 percent from the positive radiative forcing caused by human activity. However, the relative uncertainty in aerosol radiative forcing (approximately 90 percent) is much greater than that of greenhouse gases. In addition, future emissions of aerosols from human activities, and the influence of these emissions on future climate change, are not known with any certainty. Nevertheless, it can be said that, if concentrations of anthropogenic aerosols continue to decrease as they have since the 1970s, a significant offset to the effects of greenhouse gases will be reduced, opening future climate to further warming.

Land-use Change

There are a number of ways in which changes in land use can influence climate. The most direct influence is through the alteration of Earth's albedo, or surface reflectance. For example, the replacement of forest by cropland and pasture in the middle latitudes over the past several centuries has led to an increase in albedo, which in turn has led to greater reflection of incoming solar radiation in those regions. This replacement of forest by agriculture has been associated with a change in global average radiative forcing of approximately −0.2 watt per square metre since 1750. In Europe and other major agricultural regions, such land-use conversion began more than 1,000 years ago and has proceeded nearly to completion. For Europe, the negative radiative forcing due to land-use change has probably been substantial, perhaps approaching −5 watts per square metre. The influence of early land use on radiative forcing may help to explain a long period of cooling in Europe that followed a period of relatively mild conditions roughly 1,000 years ago. It is generally believed that the mild temperatures of this "medieval warm period," which was followed by a long period of cooling, rivaled those of 20th-century Europe.

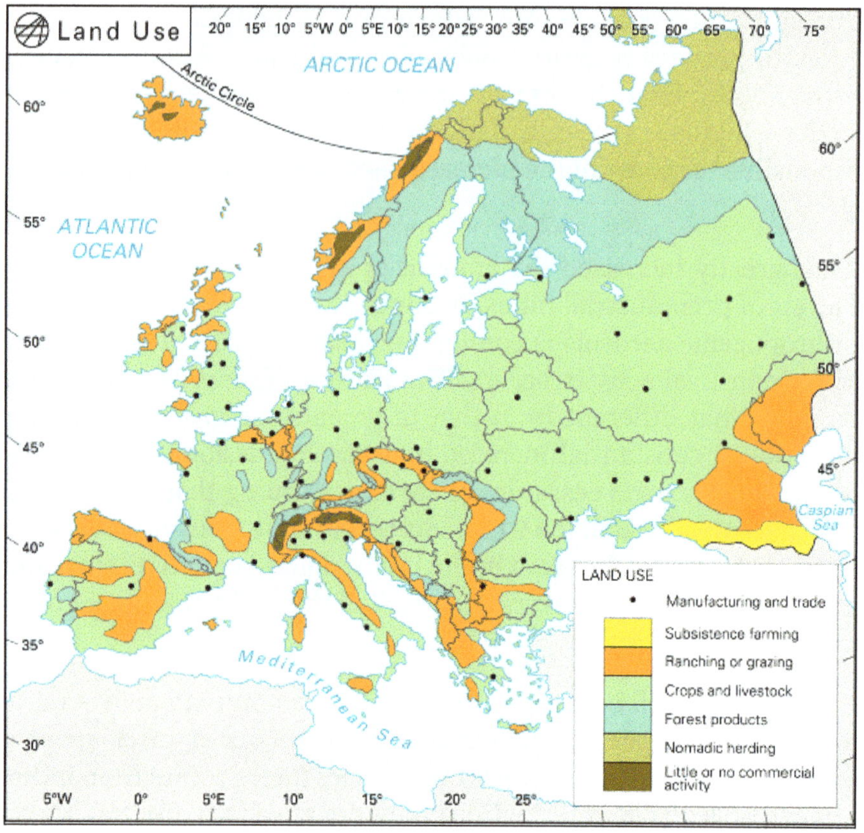

Land use in Europe.

Land-use changes can also influence climate through their influence on the exchange of heat between Earth's surface and the atmosphere. For example, vegetation helps to facilitate the evaporation of water into the atmosphere through evapotranspiration. In this process, plants take up liquid water from the soil through their root systems. Eventually this water is released through transpiration into the atmosphere, as water vapour through the stomata in leaves. While deforestation generally leads to surface cooling due to the albedo factor discussed above, the land surface may also be warmed as a result of the release of latent heat by the evapotranspiration process. The relative importance of these two factors, one exerting a cooling effect and the other a warming effect, varies by both season and region. While the albedo effect is likely to dominate in middle latitudes, especially during the period from autumn through spring, the evapotranspiration effect may dominate during the summer in the midlatitudes and year-round in the tropics. The latter case is particularly important in assessing the potential impacts of continued tropical deforestation.

The rate at which tropical regions are deforested is also relevant to the process of carbon sequestration, the long-term storage of carbon in underground cavities and biomass rather than in the atmosphere. By removing carbon from the atmosphere, carbon sequestration acts to mitigate global warming. Deforestation contributes to global warming, as fewer plants are available to take up carbon dioxide from the atmosphere. In addition, as fallen trees, shrubs, and other plants are burned or allowed to slowly decompose, they release as carbon dioxide the carbon they stored during their lifetimes. Furthermore, any land-use change that influences the amount, distribution, or type of vegetation in a region can affect the concentrations of biogenic aerosols, though the impact of such changes on climate is indirect and relatively minor.

Stratospheric Ozone Depletion

Since the 1970s the loss of ozone (O_3) from the stratosphere has led to a small amount of negative radiative forcing of the surface. This negative forcing represents a competition between two distinct effects caused by the fact that ozone absorbs solar radiation. In the first case, as ozone levels in the stratosphere are depleted, more solar radiation reaches Earth's surface. In the absence of any other influence, this rise in insolation would represent a positive radiative forcing of the surface. However, there is a second effect of ozone depletion that is related to its greenhouse properties. As the amount of ozone in the stratosphere is decreased, there is also less ozone to absorb longwave radiation emitted by Earth's surface. With less absorption of radiation by ozone, there is a corresponding decrease in the downward reemission of radiation. This second effect overwhelms the first and results in a modest negative radiative forcing of Earth's surface and a modest cooling of the lower stratosphere by approximately 0.5 °C (0.9 °F) per decade since the 1970s.

Natural Influences on Climate

There are a number of natural factors that influence Earth's climate. These factors include external influences such as explosive volcanic eruptions, natural variations in the output of the Sun, and slow changes in the configuration of Earth's orbit relative to the Sun. In addition, there are natural oscillations in Earth's climate that alter global patterns of wind circulation, precipitation, and surface temperatures. One such phenomenon is the El Niño/Southern Oscillation (ENSO), a coupled atmospheric and oceanic event that occurs in the Pacific Ocean every three to seven years. In addition, the Atlantic Multidecadal Oscillation (AMO) is a similar phenomenon that occurs over decades in the North Atlantic Ocean. Other types of oscillatory behaviour that produce dramatic shifts in climate may occur across timescales of centuries and millennia.

Volcanic Aerosols

Explosive volcanic eruptions have the potential to inject substantial amounts of sulfate aerosols into the lower stratosphere. In contrast to aerosol emissions in the lower troposphere, aerosols that enter the stratosphere may remain for several years before settling out, because of the relative absence of turbulent motions there. Consequently, aerosols from explosive volcanic eruptions have the potential to affect Earth's climate. Less-explosive eruptions, or eruptions that are less vertical in orientation, have a lower potential for substantial climate impact. Furthermore, because of large-scale circulation patterns within the stratosphere, aerosols injected within tropical regions tend to spread out over the globe, whereas aerosols injected within midlatitude and polar regions tend to remain confined to the middle and high latitudes of that hemisphere. Tropical eruptions, therefore, tend to have a greater climatic impact than eruptions occurring toward the poles. In 1991 the moderate eruption of Mount Pinatubo in the Philippines provided a peak forcing of approximately −4 watts per square metre and cooled the climate by about 0.5 °C (0.9 °F) over the following few years. By comparison, the 1815 Mount Tambora eruption in present-day Indonesia, typically implicated for the 1816 "year without a summer" in Europe and North America, is believed to have been associated with a radiative forcing of approximately −6 watts per square metre.

A column of gas and ash rising from Mount Pinatubo in the Philippines.

While in the stratosphere, volcanic sulfate aerosol actually absorbs longwave radiation emitted by Earth's surface, and absorption in the stratosphere tends to result in a cooling of the troposphere below. This vertical pattern of temperature change in the atmosphere influences the behaviour of winds in the lower atmosphere, primarily in winter. Thus, while there is essentially a global cooling effect for the first few years following an explosive volcanic eruption, changes in the winter patterns of surface winds may actually lead to warmer winters in some areas, such as Europe. Some modern examples of major eruptions include Krakatoa (Indonesia) in 1883, El Chichón (Mexico) in 1982, and Mount Pinatubo in 1991. There is also evidence that volcanic eruptions may influence other climate phenomena such as ENSO.

Variations in Solar Output

Direct measurements of solar irradiance, or solar output, have been available from satellites only since the late 1970s. These measurements show a very small peak-to-peak variation in solar irradiance (roughly 0.1 percent of the 1,366 watts per square metre received at the top of the atmosphere, for approximately 1.4 watts per square metre). However, indirect measures of solar activity are available from historical sunspot measurements dating back through the early 17th century. Attempts have been made to reconstruct graphs of solar irradiance variations from historical sunspot data by calibrating them against the measurements from modern satellites. However, since the modern measurements span only a few of the most recent 11-year solar cycles, estimates of solar output variability on 100-year and longer timescales are poorly correlated. Different assumptions regarding the relationship between the amplitudes of 11-year solar cycles and long-period solar output changes can lead to considerable differences in the resulting solar reconstructions. These differences in turn lead to fairly large uncertainty in estimating positive forcing by changes in solar irradiance since 1750. (Estimates range from 0.06 to 0.3 watt per square metre.) Even more challenging, given the lack of any modern analog, is the estimation of solar irradiance during the so-called Maunder Minimum, a period lasting from the mid-17th century to the early 18th century when very few sunspots were observed. While it is likely that solar irradiance was reduced at this time, it is difficult to calculate by how much. However, additional proxies of solar output exist that match reasonably well with the sunspot-derived records following the Maunder Minimum; these may be used as crude estimates of the solar irradiance variations.

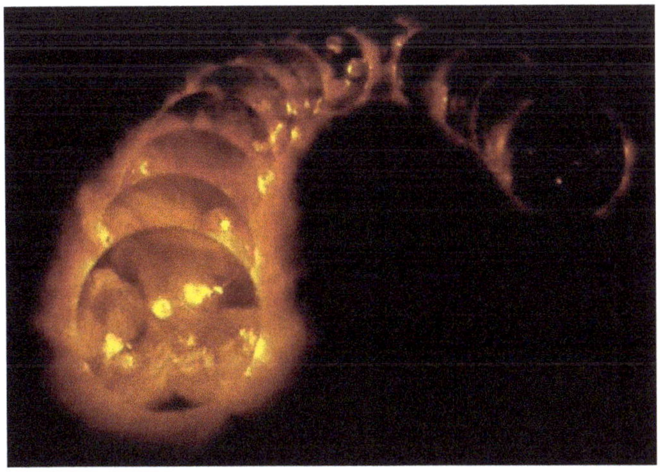

In theory it is possible to estimate solar irradiance even farther back in time, over at least the past millennium, by measuring levels of cosmogenic isotopes such as carbon-14 and beryllium-10. Cosmogenic isotopes are isotopes that are formed by interactions of cosmic rays with atomic nuclei in the atmosphere and that subsequently fall to Earth, where they can be measured in the annual layers found in ice cores. Since their production rate in the upper atmosphere is modulated by changes in solar activity, cosmogenic isotopes may be used as indirect indicators of solar irradiance. However, as with the sunspot data, there is still considerable uncertainty in the amplitude of past solar variability implied by these data.

Solar forcing also affects the photochemical reactions that manufacture ozone in the stratosphere. Through this modulation of stratospheric ozone concentrations, changes in solar irradiance (particularly in the ultraviolet portion of the electromagnetic spectrum) can modify how both shortwave and longwave radiation in the lower stratosphere are absorbed. As a result, the vertical temperature profile of the atmosphere can change, and this change can in turn influence phenomena such as the strength of the winter jet streams.

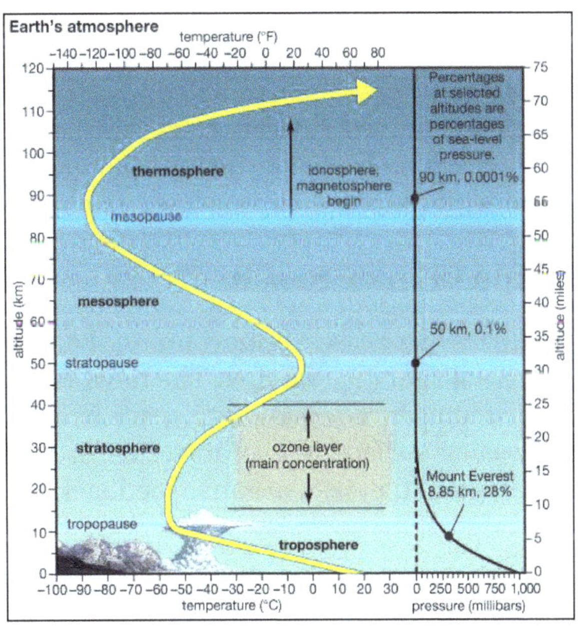

The layers of Earth's atmosphere. The yellow line shows the response of air temperature to increasing height.

Variations in Earth's Orbit

On timescales of tens of millennia, the dominant radiative forcing of Earth's climate is associated with slow variations in the geometry of Earth's orbit about the Sun. These variations include the precession of the equinoxes (that is, changes in the timing of summer and winter), occurring on a roughly 26,000-year timescale; changes in the tilt angle of Earth's rotational axis relative to the plane of Earth's orbit around the Sun, occurring on a roughly 41,000-year timescale; and changes in the eccentricity (the departure from a perfect circle) of Earth's orbit around the Sun, occurring on a roughly 100,000-year timescale. Changes in eccentricity slightly influence the mean annual solar radiation at the top of Earth's atmosphere, but the primary influence of all the orbital variations listed above is on the seasonal and latitudinal distribution of incoming solar radiation over Earth's surface. The major ice ages of the Pleistocene Epoch were closely related to the influence of these variations on summer insolation at high northern latitudes. Orbital variations thus exerted a primary control on the extent of continental ice sheets. However, Earth's orbital changes are generally believed to have had little impact on climate over the past few millennia, and so they are not considered to be significant factors in present-day climate variability.

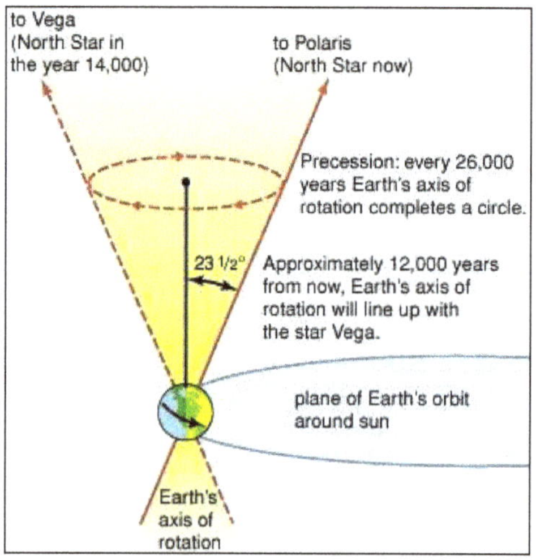

Earth's axis of rotation itself rotates, or precesses, completing one circle every 26,000 years. Consequently, Earth's North Pole points toward different stars (and sometimes toward empty space) as it travels in this circle. This precession is so slow that it is not noticeable in a person's lifetime, though astronomers must consider its effect when studying ancient sites such as Stonehenge.

Feedback Mechanisms and Climate Sensitivity

There are a number of feedback processes important to Earth's climate system and, in particular, its response to external radiative forcing. The most fundamental of these feedback mechanisms involves the loss of longwave radiation to space from the surface. Since this radiative loss increases with increasing surface temperatures according to the Stefan-Boltzmann law, it represents a stabilizing factor (that is, a negative feedback) with respect to near-surface air temperature.

Climate sensitivity can be defined as the amount of surface warming resulting from each additional watt per square metre of radiative forcing. Alternatively, it is sometimes defined as the warming that would result from a doubling of CO_2 concentrations and the associated addition of 4 watts per square metre of radiative forcing. In the absence of any additional feedbacks, climate sensitivity would be approximately 0.25 °C (0.45 °F) for each additional watt per square metre of radiative forcing. Stated alternatively, if the CO_2 concentration of the atmosphere present at the start of the industrial age (280 ppm) were doubled (to 560 ppm), the resulting additional 4 watts per square metre of radiative forcing would translate into a 1 °C (1.8 °F) increase in air temperature. However, there are additional feedbacks that exert a destabilizing, rather than stabilizing, influence, and these feedbacks tend to increase the sensitivity of climate to somewhere between 0.5 and 1.0 °C (0.9 and 1.8 °F) for each additional watt per square metre of radiative forcing.

Water Vapour Feedback

Unlike concentrations of other greenhouse gases, the concentration of water vapour in the atmosphere cannot freely vary. Instead, it is determined by the temperature of the lower atmosphere and surface through a physical relationship known as the Clausius-Clapeyron equation, named for 19th-century German physicist Rudolf Clausius and 19th-century French engineer Émile Clapeyron. Under the assumption that there is a liquid water surface in equilibrium with the atmosphere, this relationship indicates that an increase in the capacity of air to hold water vapour is a function of increasing temperature of that volume of air. This assumption is relatively good over the oceans, where water is plentiful, but not over the continents. For this reason the relative humidity (the percent of water vapour the air contains relative to its capacity) is approximately 100 percent over ocean regions and much lower over continental regions (approaching 0 percent in arid regions). Not surprisingly, the average relative humidity of Earth's lower atmosphere is similar to the fraction of Earth's surface covered by the oceans (that is, roughly 70 percent). This quantity is expected to remain approximately constant as Earth warms or cools. Slight changes to global relative humidity may result from human land-use modification, such as tropical deforestation and irrigation, which can affect the relative humidity over land areas up to regional scales.

The amount of water vapour in the atmosphere will rise as the temperature of the atmosphere rises. Since water vapour is a very potent greenhouse gas, even more potent than CO_2, the net greenhouse effect actually becomes stronger as the surface warms, which leads to even greater warming. This positive feedback is known as the "water vapour feedback." It is the primary reason that climate sensitivity is substantially greater than the previously stated theoretical value of 0.25 °C (0.45 °F) for each increase of 1 watt per square metre of radiative forcing.

Cloud Feedbacks

It is generally believed that as Earth's surface warms and the atmosphere's water vapour content increases, global cloud cover increases. However, the effects on near-surface air temperatures are complicated. In the case of low clouds, such as marine stratus clouds, the dominant radiative feature of the cloud is its albedo. Here any increase in low cloud cover acts in much the same way as an increase in surface ice cover: more incoming solar radiation is reflected and Earth's surface cools. On the other hand, high clouds, such as the towering cumulus clouds that extend up to the boundary between the troposphere and stratosphere, have a quite different impact on the surface radiation balance. The tops of cumulus clouds are considerably higher in the atmosphere and colder than their undersides. Cumulus cloud tops emit less longwave radiation out to space than the warmer cloud bottoms emit downward toward the surface. The end result of the formation of high cumulus clouds is greater warming at the surface.

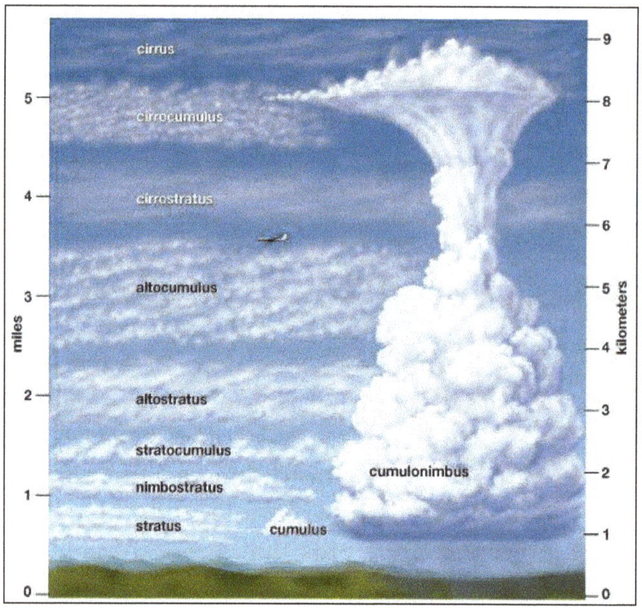

Different types of clouds form at different heights.

The net feedback of clouds on rising surface temperatures is therefore somewhat uncertain. It represents a competition between the impacts of high and low clouds, and the balance is difficult to determine. Nonetheless, most estimates indicate that clouds on the whole represent a positive feedback and thus additional warming.

Ice Albedo Feedback

Another important positive climate feedback is the so-called ice albedo feedback. This feedback arises from the simple fact that ice is more reflective (that is, has a higher albedo) than land or water surfaces. Therefore, as global ice cover decreases, the reflectivity of Earth's surface decreases, more incoming solar radiation is absorbed by the surface, and the surface warms. This feedback is considerably more important when there is relatively extensive global ice cover, such as during the height of the last ice age, roughly 25,000 years ago. On a global scale the importance of ice albedo feedback decreases as Earth's surface warms and there is relatively less ice available to be melted.

Carbon Cycle Feedbacks

Another important set of climate feedbacks involves the global carbon cycle. In particular, the two main reservoirs of carbon in the climate system are the oceans and the terrestrial biosphere. These reservoirs have historically taken up large amounts of anthropogenic CO_2 emissions. Roughly 50–70 percent is removed by the oceans, whereas the remainder is taken up by the terrestrial biosphere. Global warming, however, could decrease the capacity of these reservoirs to sequester atmospheric CO_2. Reductions in the rate of carbon uptake by these reservoirs would increase the pace of CO_2 buildup in the atmosphere and represent yet another possible positive feedback to increased greenhouse gas concentrations.

In the world's oceans, this feedback effect might take several paths. First, as surface waters warm, they would hold less dissolved CO_2. Second, if more CO_2 were added to the atmosphere and taken up by the oceans, bicarbonate ions (HCO_3–) would multiply and ocean acidity would increase. Since calcium carbonate ($CaCO_3$) is broken down by acidic solutions, rising acidity would threaten ocean-dwelling fauna that incorporate $CaCO_3$ into their skeletons or shells. As it becomes increasingly difficult for these organisms to absorb oceanic carbon, there would be a corresponding decrease in the efficiency of the biological pump that helps to maintain the oceans as a carbon sink (as described in the section Carbon dioxide). Third, rising surface temperatures might lead to a slowdown in the so-called thermohaline circulation, a global pattern of oceanic flow that partly drives the sinking of surface waters near the poles and is responsible for much of the burial of carbon in the deep ocean. A slowdown in this flow due to an influx of melting fresh water into what are normally saltwater conditions might also cause the solubility pump, which transfers CO_2 from shallow to deeper waters, to become less efficient. Indeed, it is predicted that if global warming continued to a certain point, the oceans would cease to be a net sink of CO_2 and would become a net source.

As large sections of tropical forest are lost because of the warming and drying of regions such as Amazonia, the overall capacity of plants to sequester atmospheric CO_2 would be reduced. As a result, the terrestrial biosphere, though currently a carbon sink, would become a carbon source. Ambient temperature is a significant factor affecting the pace of photosynthesis in plants, and many plant species that are well adapted to their local climatic conditions have maximized their photosynthetic rates. As temperatures increase and conditions begin to exceed the optimal temperature range for both photosynthesis and soil respiration, the rate of photosynthesis would decline. As dead plants decompose, microbial metabolic activity (a CO_2 source) would increase and would eventually outpace photosynthesis.

Under sufficient global warming conditions, methane sinks in the oceans and terrestrial biosphere also might become methane sources. Annual emissions of methane by wetlands might either increase or decrease, depending on temperatures and input of nutrients, and it is possible that wetlands could switch from source to sink. There is also the potential for increased methane release as a result of the warming of Arctic permafrost (on land) and further methane release at the continental margins of the oceans (a few hundred metres below sea level). The current average atmospheric methane concentration of 1,750 ppb is equivalent to 3.5 gigatons (3.5 billion tons) of carbon. There are at least 400 gigatons of carbon equivalent stored in Arctic permafrost and as much as 10,000 gigatons (10 trillion tons) of carbon equivalent trapped on the continental margins of the oceans in a hydrated crystalline form known as clathrate. It is believed that some fraction of this trapped methane could become unstable with additional warming, although the amount and rate of potential emission remain highly uncertain.

Climate Research

Modern research into climatic variation and change is based on a variety of empirical and theoretical lines of inquiry. One line of inquiry is the analysis of data that record changes in atmosphere, oceans, and climate from roughly 1850 to the present. In a second line of inquiry, information describing paleoclimatic changes is gathered from "proxy" or indirect, sources such as ocean and lake sediments, pollen grains, corals, ice cores, and tree rings. Finally, a variety of theoretical models can be used to investigate the behaviour of Earth's climate under different conditions.

Modern Observations

Although a limited regional subset of land-based records is available from the 17th and 18th centuries, instrumental measurements of key climate variables have been collected systematically and at global scales since the mid-19th to early 20th century. These data include measurements of surface temperature on land and at sea, atmospheric pressure at sea level, precipitation over continents and oceans, sea ice extents, surface winds, humidity, and tides. Such records are the most reliable of all available climate data, since they are precisely dated and are based on well-understood instruments and physical principles. Corrections must be made for uncertainties in the data (for instance, gaps in the observational record, particularly during earlier years) and for systematic errors (such as an "urban heat island" bias in temperature measurements made on land).

Since the mid-20th century a variety of upper-air observations have become available (for example, of temperature, humidity, and winds), allowing climatic conditions to be characterized from the ground upward through the upper troposphere and lower stratosphere. Since the 1970s these data have been supplemented by polar-orbiting and geostationary satellites and by platforms in the oceans that gauge temperature, salinity, and other properties of seawater. Attempts have been made to fill the gaps in early measurements by using various statistical techniques and "backward prediction" models and by assimilating available observations into numerical weather prediction models. These techniques seek to estimate meteorological observations or atmospheric variables (such as relative humidity) that have been poorly measured in the past.

Modern measurements of greenhouse gas concentrations began with an investigation of atmospheric carbon dioxide (CO_2) concentrations by American climate scientist Charles Keeling at the summit of Mauna Loa in Hawaii in 1958. Keeling's findings indicated that CO_2 concentrations were steadily rising in association with the combustion of fossil fuels, and they also yielded the famous "Keeling curve," a graph in which the longer-term rising trend is superimposed on small oscillations related to seasonal variations in the uptake and release of CO_2 from photosynthesis and respiration in the terrestrial biosphere. Keeling's measurements at Mauna Loa apply primarily to the Northern Hemisphere.

Taking into account the uncertainties, the instrumental climate record indicates substantial trends since the end of the 19th century consistent with a warming Earth. These trends include a rise in global surface temperature of 0.9 °C (1.5 °F) between 1880 and 2012, an associated elevation of global sea level of 19–21 cm (7.5–8.3 inches) between 1901 and 2010, and a decrease in snow cover in the Northern Hemisphere of approximately 1.5 million square km (580,000 square miles). Records of average global temperatures kept by the World Meteorological Organization (WMO) indicate that the years 1998, 2005, and 2010 are statistically tied with one another as the warmest years since modern record keeping began in 1880; the WMO also noted that the decade 2001–10 was the warmest decade since 1880. Increases in global sea level are attributed to a combination of seawater expansion due to ocean heating and freshwater runoff caused by the melting of terrestrial ice. Reductions in snow cover are the result of warmer temperatures favouring a steadily shrinking winter season.

Climate data collected during the first two decades of the 21st century reveal that surface warming between 2005 and 2014 proceeded slightly more slowly than was expected from the effect of greenhouse gas increases alone. This fact was sometimes used to suggest that global warming had stopped or that it experienced a "hiatus" or "pause." In reality, this phenomenon appears to have been influenced by several factors, none of which, however, implies that global warming stopped during this period or that global warming would not continue in the future. One factor was the increased burial of heat beneath the ocean surface by strong trade winds, a process assisted by La Niña conditions. The effects of La Niña manifest in the form of cooling surface waters along the western coast of South America. As a result, warming at the ocean surface was reduced, but the accumulation of heat in other parts of the ocean occurred at an accelerated rate. Another factor cited by climatologists was a small but potentially important increase in aerosols from volcanic activity, which may have blocked a small portion of incoming solar radiation and which were accompanied by a small reduction in solar output during the period. These factors, along with natural decades-long oscillations in the climate system, may have masked a portion of the greenhouse warming. (However, climatologists point out that these natural climate cycles are expected to add to greenhouse warming in the future when the oscillations eventually reverse direction.) For these reasons many scientists believe that it is an error to call this slowdown in detectable surface warming a "hiatus" or a "pause."

Prehistorical Climate Records

In order to reconstruct climate changes that occurred prior to about the mid-19th century, it is necessary to use "proxy" measurements—that is, records of other natural phenomena that indirectly measure various climate conditions. Some proxies, such as most sediment cores and pollen records, glacial moraine evidence, and geothermal borehole temperature profiles, are coarsely resolved or dated and thus are only useful for describing climate changes on long timescales. Other proxies, such as growth rings from trees or oxygen isotopes from corals and ice cores, can provide a record of yearly or even seasonal climate changes.

The data from these proxies should be calibrated to known physical principles or related statistically to the records collected by modern instruments, such as satellites. Networks of proxy data can then be used to infer patterns of change in climate variables, such as the behaviour of surface temperature over time and geography. Yearly reconstructions of climate variables are possible over the past 1,000 to 2,000 years using annually dated proxy records, but reconstructions farther back in time are generally based on more coarsely resolved evidence such as ocean sediments and pollen records. For these, records of conditions can be reconstructed only on timescales of hundreds or thousands of years. In addition, since relatively few long-term proxy records are available for the Southern Hemisphere, most reconstructions focus on the Northern Hemisphere.

The various proxy-based reconstructions of the average surface temperature of the Northern Hemisphere differ in their details. These differences are the result of uncertainties implicit in the proxy data themselves and also of differences in the statistical methods used to relate the proxy data to surface temperature. Nevertheless, all studies as reviewed in the IPCC's Fourth Assessment Report (AR4), which was published in 2007, indicate that the average surface temperature since about 1950 is higher than at any time during the previous 1,000 years.

Theoretical Climate Models

Theoretical models of Earth's climate system can be used to investigate the response of climate to external radiative forcing as well as its own internal variability. Two or more models that focus on different physical processes may be coupled or linked together through a common feature, such as geographic location. Climate models vary considerably in their degree of complexity. The simplest models of energy balance describe Earth's surface as a globally uniform layer whose temperature is determined by a balance of incoming and outgoing shortwave and longwave radiation. These simple models may also consider the effects of greenhouse gases. At the other end of the spectrum are fully coupled, three-dimensional, global climate models. These are complex models that solve for radiative balance; for laws of motion governing the atmosphere, ocean, and ice; and for exchanges of energy and momentum within and between the different components of the climate. In some cases, theoretical climate models also include an interactive representation of Earth's biosphere and carbon cycle.

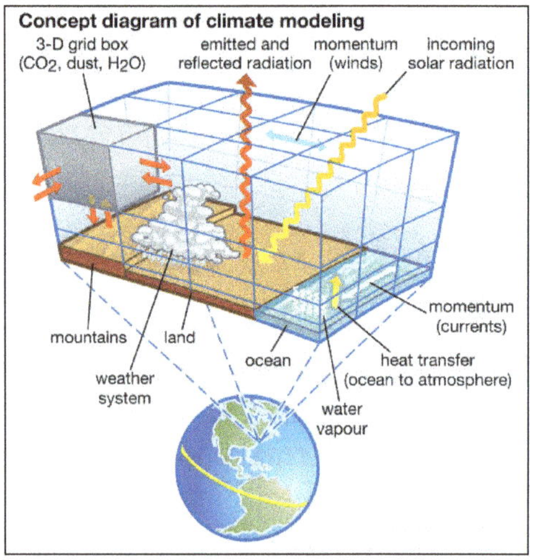

To understand and explain the complex behaviour of Earth's climate, modern climate models incorporate several variables that stand in for materials passing through Earth's atmosphere and oceans and the forces that affect them.

Even the most-detailed climate models cannot resolve all the processes that are important in the atmosphere and ocean. Most climate models are designed to gauge the behaviour of a number of physical variables over space and time, and they often artificially divide Earth's surface into a grid of many equal-sized "cells". Each cell may neatly correspond to some physical process (such as summer near-surface air temperature) or other variable (such as land-use type), and it may be assigned a relatively straightforward value. So-called "sub-grid-scale" processes, such as those of clouds, are too small to be captured by the relatively coarse spacing of the individual grid cells. Instead, such processes must be represented through a statistical process that relates the properties of the atmosphere and ocean. For example, the average fraction of cloud cover over a hypothetical "grid box" (that is, a representative volume of air or water in the model) can be estimated from the average relative humidity and the vertical temperature profile of the grid cell. Variations in the behaviour of different coupled climate models arise in large part from differences in the ways sub-grid-scale processes are mathematically expressed.

Despite these required simplifications, many theoretical climate models perform remarkably well when reproducing basic features of the atmosphere, such as the behaviour of midlatitude jet streams or Hadley cell circulation. The models also adequately reproduce important features of the oceans, such as the Gulf Stream. In addition, models are becoming better able to reproduce the main patterns of internal climate variability, such as those of El Niño/Southern Oscillation (ENSO). Consequently, periodically recurring events—such as ENSO and other interactions between the atmosphere and ocean currents—are being modeled with growing confidence.

Climate models have been tested in their ability to reproduce observed changes in response to radiative forcing. In 1988 a team at NASA's Goddard Institute for Space Studies in New York City used a fairly primitive climate model to predict warming patterns that might occur in response to three different scenarios of anthropogenic radiative forcing. Warming patterns were forecast for subsequent decades. Of the three scenarios, the middle one, which corresponds most closely to actual historical carbon emissions, comes closest to matching the observed warming of roughly 0.5 °C (0.9 °F) that has taken place since then. The NASA team also used a climate model to successfully predict that global mean surface temperatures would cool by about 0.5 °C for one to two years after the 1991 eruption of Mount Pinatubo in the Philippines.

More recently, so-called "detection and attribution" studies have been performed. These studies compare predicted changes in near-surface air temperature and other climate variables with patterns of change that have been observed for the past one to two centuries. The simulations have shown that the observed patterns of warming of Earth's surface and upper oceans, as well as changes in other climate phenomena such as prevailing winds and precipitation patterns, are consistent with the effects of an anthropogenic influence predicted by the climate models. In addition, climate model simulations have shown success in reproducing the magnitude and the spatial pattern of cooling in the Northern Hemisphere between roughly 1400 and 1850—during the Little Ice Age, which appears to have resulted from a combination of lowered solar output and heightened explosive volcanic activity.

Potential Effects of Global Warming

The path of future climate change will depend on what courses of action are taken by society—in particular the emission of greenhouse gases from the burning of fossil fuels. A range of alternative emissions scenarios known as representative concentration pathways (RCPs) were proposed by the IPCC in the Fifth Assessment Report (AR5), which was published in 2014, to examine potential future climate changes. The scenarios depend on various assumptions concerning future rates of human population growth, economic development, energy demand, technological advancement, and other factors. Unlike the scenarios used in previous IPCC assessments, the AR5 RCPs explicitly account for climate change mitigation efforts.

Table: Projected range of sea-level rise by climate change scenario.

Scenario	Temperature change (°C) in 2090–99 relative to 1980–99	Sea-level rise (m) in 2090–99 relative to 1980–99
B1	1.1–2.9	0.18–0.38
A1T	1.4–3.8	0.20–0.45
B2	1.4–3.8	0.20–0.43
A1B	1.7–4.4	0.21–0.48
A2	2.0–5.4	0.23–0.51
A1Fl	2.4–6.4	0.26–0.59

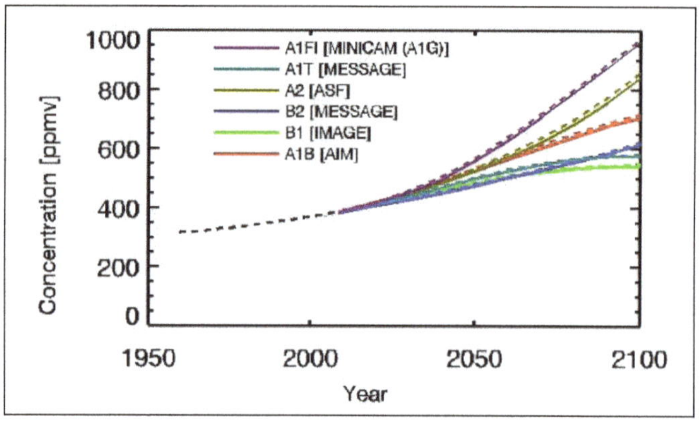

Carbon dioxide: global warming scenarios Graph of the predicted increase in the concentration of carbon dioxide (CO_2) in Earth's atmosphere according to a series of climate change scenarios that assume different levels of economic development, population growth, and fossil fuel use. The results of each scenario in the IPCC's Fourth Assessment Report (2007) are depicted in the graph.

The AR5 scenario with the smallest increases in greenhouse gases is RCP 2.6, which denotes the net radiative forcing by 2100 in watts per square metre (a doubling of CO2 concentrations from preindustrial values of 280 ppm to 560 ppm represents roughly 3.7 watts per square metre). RCP 2.6 assumes substantial improvements in energy efficiency, a rapid transition away from fossil fuel energy, and a global population that peaks at roughly nine billion people in the 21st century. In that scenario CO_2 concentrations remain below 450 ppm and actually fall toward the end of the century (to about 420 ppm) as a result of widespread deployment of carbon-capture technology.

Scenario RCP 8.5, by contrast, might be described as "business as usual." It reflects the assumption of an energy-intensive global economy, high population growth, and a reduced rate of technological development. CO_2 concentrations are more than three times greater than preindustrial levels (roughly 936 ppm) by 2100 and continue to grow thereafter. RCP 4.5 and RCP 6.0 envision intermediate policy choices, resulting in stabilization by 2100 of CO_2 concentrations at 538 and 670 ppm, respectively. In all those scenarios, the cooling effect of industrial pollutants such as sulfate particulates, which have masked some of the past century's warming, is assumed to decline to near zero by 2100 because of policies restricting their industrial production.

Simulations of Future Climate Change

The differences between the various simulations arise from disparities between the various climate models used and from assumptions made by each emission scenario. For example, best estimates of the predicted increases in global surface temperature between the years 2000 and 2100 range from about 0.3 to 4.8 °C (0.5 to 8.6 °F), depending on which emission scenario is assumed and which climate model is used. Relative to preindustrial (i.e., 1750–1800) temperatures, these estimates reflect an overall warming of the globe of 1.4 to 5.0 °C (2.5 to 9.0 °F). These projections are conservative in that they do not take into account potential positive carbon cycle feedbacks. Only the lower-end emissions scenario RCP 2.6 has a reasonable chance (roughly 50 percent) of holding additional global surface warming by 2100 to less than 2.0 °C (3.6 °F)—a level considered by many scientists to be the threshold above which pervasive and extreme climatic effects will occur.

Patterns of Warming

The greatest increase in near-surface air temperature is projected to occur over the polar region of the Northern Hemisphere because of the melting of sea ice and the associated reduction in surface albedo. Greater warming is predicted over land areas than over the ocean. Largely due to the delayed warming of the oceans and their greater specific heat, the Northern Hemisphere—with less than 40 percent of its surface area covered by water—is expected to warm faster than the Southern Hemisphere. Some of the regional variation in predicted warming is expected to arise from changes to wind patterns and ocean currents in response to surface warming. For example, the warming of the region of the North Atlantic Ocean just south of Greenland is expected to be slight. This anomaly is projected to arise from a weakening of warm northward ocean currents combined with a shift in the jet stream that will bring colder polar air masses to the region.

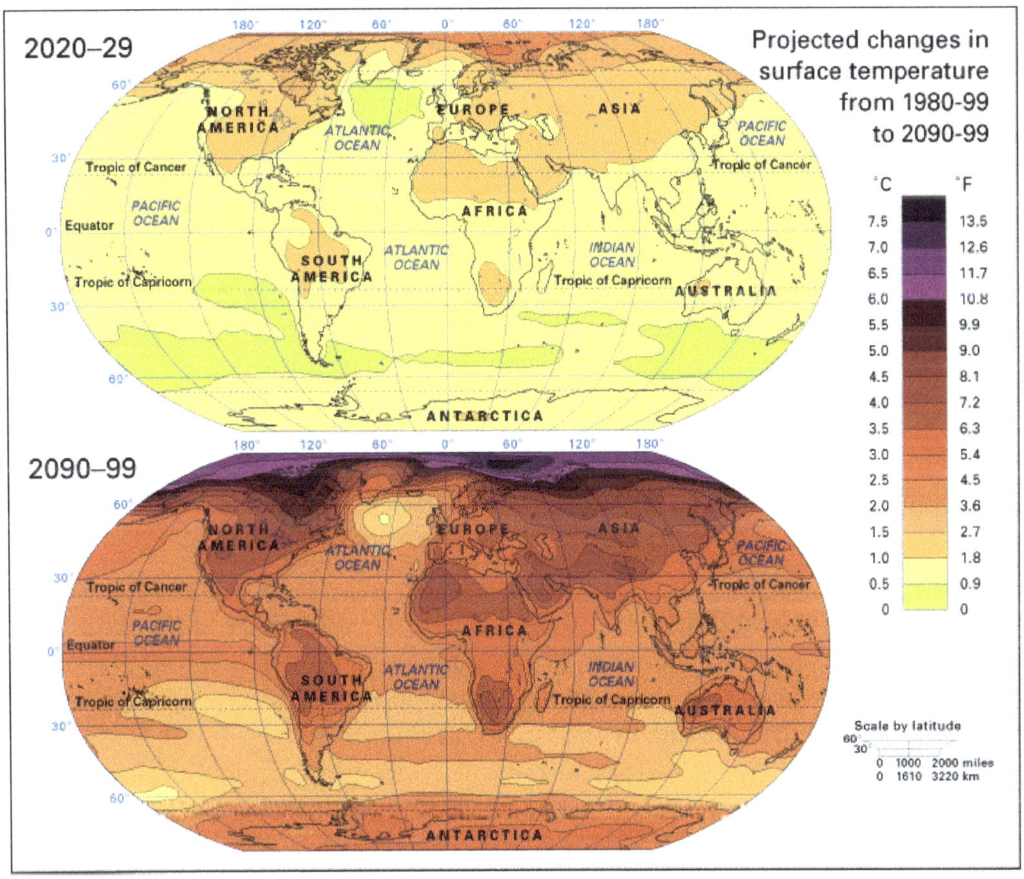

Projected changes in mean surface temperatures by the late 21st century according to the A1B climate change scenario. All values for the period 2090–99 are shown relative to the mean temperature values for the period 1980–99.

Precipitation Patterns

The climate changes associated with global warming are also projected to lead to changes in precipitation patterns across the globe. Increased precipitation is predicted in the polar and subpolar regions, whereas decreased precipitation is projected for the middle latitudes of both hemispheres as a result of the expected poleward shift in the jet streams. Whereas precipitation near the Equator is predicted to increase, it is thought that rainfall in the subtropics will decrease. Both phenomena are associated with a forecasted strengthening of the tropical Hadley cell pattern of atmospheric circulation.

Changes in precipitation patterns are expected to increase the chances of both drought and flood conditions in many areas. Decreased summer precipitation in North America, Europe, and Africa, combined with greater rates of evaporation due to warming surface temperatures, is projected to lead to decreased soil moisture and drought in many regions. Furthermore, since anthropogenic climate change will likely lead to a more vigorous hydrologic cycle with greater rates of both evaporation and precipitation, there will be a greater probability for intense precipitation and flooding in many regions.

Regional Predictions

Regional predictions of future climate change remain limited by uncertainties in how the precise patterns of atmospheric winds and ocean currents will vary with increased surface warming. For example, some uncertainty remains in how the frequency and magnitude of El Niño/ Southern Oscillation (ENSO) events will adjust to climate change. Since ENSO is one of the most prominent sources of interannual variations in regional patterns of precipitation and temperature, any uncertainty in how it will change implies a corresponding uncertainty in certain regional patterns of climate change. For example, increased El Niño activity would likely lead to more winter precipitation in some regions, such as the desert southwest of the United States. This might offset the drought predicted for those regions, but at the same time it might lead to less precipitation in other regions. Rising winter precipitation in the desert southwest of the United States might exacerbate drought conditions in locations as far away as South Africa.

Ice Melt and Sea Level Rise

A warming climate holds important implications for other aspects of the global environment. Because of the slow process of heat diffusion in water, the world's oceans are likely to continue to warm for several centuries in response to increases in greenhouse concentrations that have taken place so far. The combination of seawater's thermal expansion associated with this warming and the melting of mountain glaciers is predicted to lead to an increase in global sea level of 0.45–0.82 metre (1.4–2.7 feet) by 2100 under the RCP 8.5 emissions scenario. However, the actual rise in sea level could be considerably greater than this. It is probable that the continued warming of Greenland will cause its ice sheet to melt at accelerated rates. In addition, this level of surface warming may also melt the ice sheet of West Antarctica. Paleoclimatic evidence suggests that an additional 2 °C (3.6 °F) of warming could lead to the ultimate destruction of the Greenland Ice Sheet, an event that would add another 5 to 6 metres (16 to 20 feet) to predicted sea level rise. Such an increase would submerge a substantial number of islands and lowland regions. Coastal lowland regions vulnerable to sea level rise include substantial parts of the U.S. Gulf Coast and Eastern Seaboard (including roughly the lower third of Florida), much of the Netherlands and Belgium (two of the European Low Countries), and heavily populated tropical areas such as Bangladesh. In addition, many of the world's major cities—such as Tokyo, New York, Mumbai, Shanghai, and Dhaka—are located in lowland regions vulnerable to rising sea levels. With the loss of the West Antarctic ice sheet, additional sea level rise would approach 10.5 metres (34 feet).

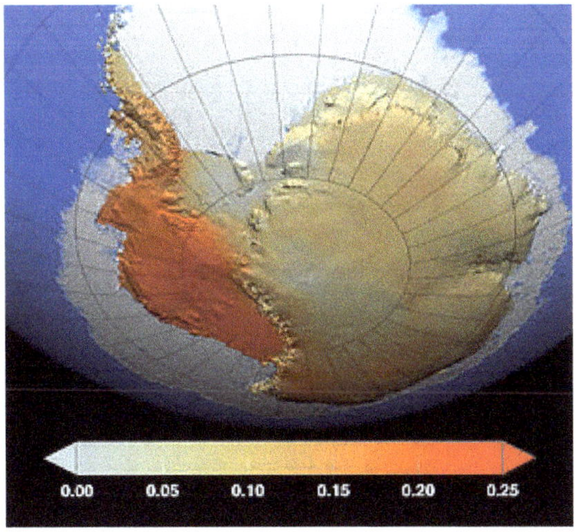

NASA image showing locations on Antarctica where temperatures had increased between 1959 and 2009. Red represents areas where temperatures had increased the most over the period, particularly in West Antarctica, while dark blue represents areas with a lesser degree of warming. Temperature changes are measured in degrees Celsius. GSFC Scientific Visualization Studio/NASA

While the current generation of models predicts that such global sea level changes might take several centuries to occur, it is possible that the rate could accelerate as a result of processes that tend to hasten the collapse of ice sheets. One such process is the development of moulins—large vertical shafts in the ice that allow surface meltwater to penetrate to the base of the ice sheet. A second process involves the vast ice shelves off Antarctica that buttress the grounded continental ice sheet of Antarctica's interior. If those ice shelves collapse, the continental ice sheet could become unstable, slide rapidly toward the ocean, and melt, thereby further increasing mean sea level. Thus far, neither process has been incorporated into the theoretical models used to predict sea level rise.

Ocean Circulation Changes

Another possible consequence of global warming is a decrease in the global ocean circulation system known as the "thermohaline circulation" or "great ocean conveyor belt." This system involves the sinking of cold saline waters in the subpolar regions of the oceans, an action that helps to drive warmer surface waters poleward from the subtropics. As a result of this process, a warming influence is carried to Iceland and the coastal regions of Europe that moderates the climate in those regions. Some scientists believe that global warming could shut down this ocean current system by creating an influx of fresh water from melting ice sheets and glaciers into the subpolar North Atlantic Ocean. Since fresh water is less dense than saline water, a significant intrusion of fresh water would lower the density of the surface waters and thus inhibit the sinking motion that drives the large-scale thermohaline circulation. It has also been speculated that, as a consequence of large-scale surface warming, such changes could even trigger colder conditions in regions surrounding the North Atlantic. Experiments with modern climate models suggest that such an event would be unlikely. Instead, a moderate weakening of the thermohaline circulation might occur that would lead to a dampening of surface warming—rather than actual cooling—in the higher latitudes of the North Atlantic Ocean.

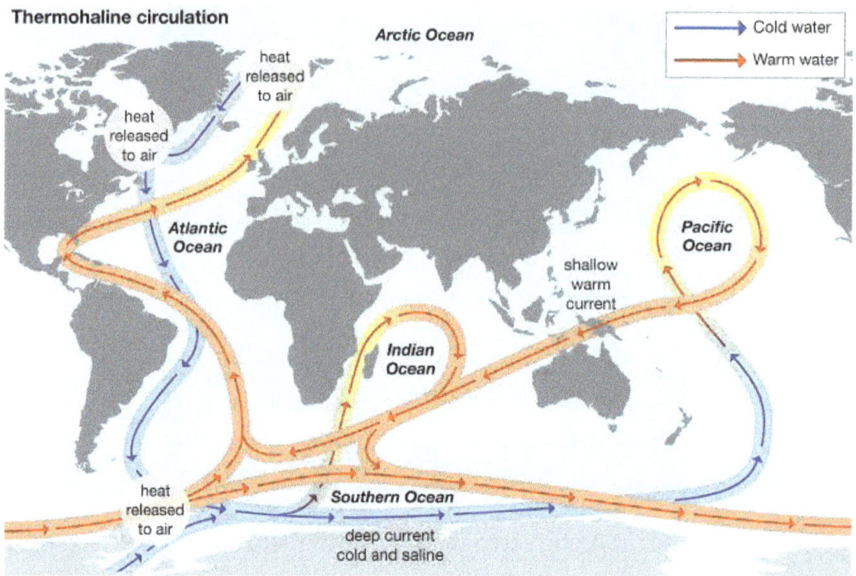

Thermohaline circulation transports and mixes the water of the oceans. In the process it transports heat, which influences regional climate patterns. The density of seawater is determined by the temperature and salinity of a volume of seawater at a particular location. The difference in density between one location and another drives the thermohaline circulation.

Tropical Cyclones

One of the more controversial topics in the science of climate change involves the impact of global warming on tropical cyclone activity. It appears likely that rising tropical ocean temperatures associated with global warming will lead to an increase in the intensity (and the associated destructive potential) of tropical cyclones. In the Atlantic a close relationship has been observed between rising ocean temperatures and a rise in the strength of hurricanes. Trends in the intensities of tropical cyclones in other regions, such as in the tropical Pacific and Indian oceans, are more uncertain due to a paucity of reliable long-term measurements.

While the warming of oceans favours increased tropical cyclone intensities, it is unclear to what extent rising temperatures affect the number of tropical cyclones that occur each year. Other factors, such as wind shear, could play a role. If climate change increases the amount of wind shear—a factor that discourages the formation of tropical cyclones—in regions where such storms tend to form, it might partially mitigate the impact of warmer temperatures. On the other hand, changes in atmospheric winds are themselves uncertain—because of, for example, uncertainties in how climate change will affect ENSO.

Environmental Consequences of Global Warming

Global warming and climate change have the potential to alter biological systems. More specifically, changes to near-surface air temperatures will likely influence ecosystem functioning and thus the biodiversity of plants, animals, and other forms of life. The current geographic ranges of plant and animal species have been established by adaptation to long-term seasonal climate patterns. As global warming alters these patterns on timescales considerably shorter than those that arose in the past from natural climate variability, relatively sudden climatic changes may challenge the natural adaptive capacity of many species.

A large fraction of plant and animal species are likely to be at an increased risk of extinction if global average surface temperatures rise another 1.5 to 2.5 °C (2.7 to 4.5 °F) by the year 2100. Species loss estimates climb to as much as 40 percent for a warming in excess of 4.5 °C (8.1 °F)—a level that could be reached in the IPCC's higher emissions scenarios. A 40 percent extinction rate would likely lead to major changes in the food webs within ecosystems and have a destructive impact on ecosystem function.

Surface warming in temperate regions is likely to lead changes in various seasonal processes—for instance, earlier leaf production by trees, earlier greening of vegetation, altered timing of egg laying and hatching, and shifts in the seasonal migration patterns of birds, fishes, and other migratory animals. In high-latitude ecosystems, changes in the seasonal patterns of sea ice threaten predators such as polar bears and walruses; both species rely on broken sea ice for their hunting activities. Also in the high latitudes, a combination of warming waters, decreased sea ice, and changes in ocean salinity and circulation is likely to lead to reductions or redistributions in populations of algae and plankton. As a result, fish and other organisms that forage upon algae and plankton may be threatened. On land, rising temperatures and changes in precipitation patterns and drought frequencies are likely to alter patterns of disturbance by fires and pests.

Numerous ecologists, conservation biologists, and other scientists studying climate warn that rising surface temperatures will bring about an increased extinction risk. In 2015 one study that examined 130 extinction models developed in previous studies predicted that 5.2 percent of species would be lost with a rise in average temperatures of 2 °C (3.6 °F) above temperature benchmarks from before the onset of the Industrial Revolution. The study also predicted that 16 percent of Earth's species would be lost if surface warming increased to about 4.3 °C (7.7 °F) above preindustrial temperature benchmarks.

Other likely impacts on the environment include the destruction of many coastal wetlands, salt marshes, and mangrove swamps as a result of rising sea levels and the loss of certain rare and fragile habitats that are often home to specialist species that are unable to thrive in other environments. For example, certain amphibians limited to isolated tropical cloud forests either have become extinct already or are under serious threat of extinction. Cloud forests—tropical forests that depend on persistent condensation of moisture in the air—are disappearing as optimal condensation levels move to higher elevations in response to warming temperatures in the lower atmosphere.

In many cases a combination of stresses caused by climate change as well as human activity represents a considerably greater threat than either climatic stresses or nonclimatic stresses alone. A particularly important example is coral reefs, which contain much of the ocean's biodiversity. Rising ocean temperatures increase the tendency for coral bleaching (a condition where zooxanthellae, or yellow-green algae, living in symbiosis with coral either lose their pigments or abandon the coral polyps altogether), and they also raise the likelihood of greater physical damage by progressively more destructive tropical cyclones. In many areas coral is also under stress from increased ocean acidification, marine pollution, runoff from agricultural fertilizer, and physical damage by boat anchors and dredging.

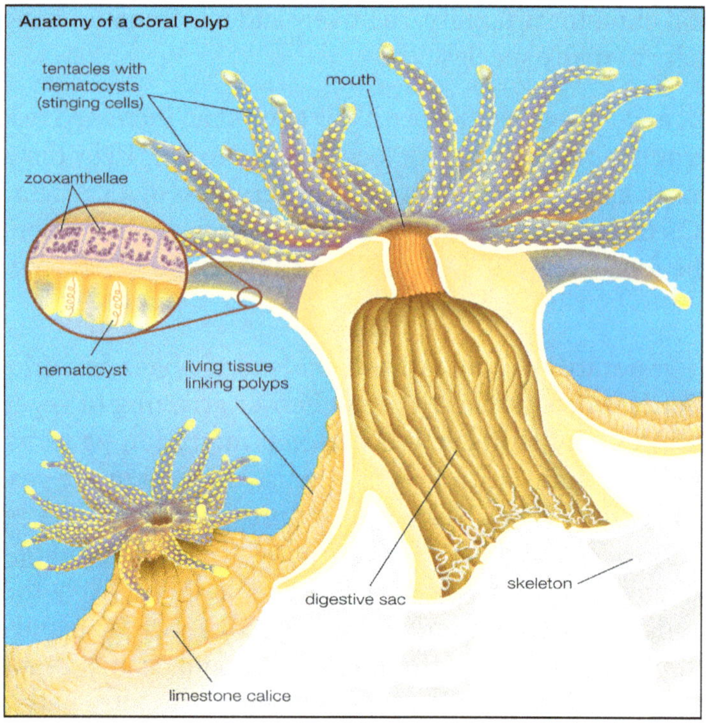

Cross section of a generalized coral polyp.

Another example of how climate and nonclimatic stresses combine is illustrated by the threat to migratory animals. As these animals attempt to relocate to regions with more favourable climate conditions, they are likely to encounter impediments such as highways, walls, artificial waterways, and other man-made structures.

Warmer temperatures are also likely to affect the spread of infectious diseases, since the geographic ranges of carriers, such as insects and rodents, are often limited by climatic conditions. Warmer winter conditions in New York in 1999, for example, appear to have facilitated an outbreak of West Nile virus, whereas the lack of killing frosts in New Orleans during the early 1990s led to an explosion of disease-carrying mosquitoes and cockroaches. Warmer winters in the Korean peninsula and southern Europe have allowed the spread of the Anopheles mosquito, which carries the malaria parasite, whereas warmer conditions in Scandinavia in recent years have allowed for the northward advance of encephalitis.

Anopheles mosquito, carrier of the malarial parasite.

In the southwestern United States, alternations between drought and flooding related in part to the ENSO phenomenon have created conditions favourable for the spread of hantaviruses by rodents. The spread of mosquito-borne Rift Valley fever in equatorial East Africa has also been related to wet conditions in the region associated with ENSO. Severe weather conditions conducive to rodents or insects have been implicated in infectious disease outbreaks—for instance, the outbreaks of cholera and leptospirosis that occurred after Hurricane Mitch struck Central America in 1998. Global warming could therefore affect the spread of infectious disease through its influence on ENSO or on severe weather conditions.

Socioeconomic Consequences of Global Warming

Socioeconomic impacts of global warming could be substantial, depending on the actual temperature increases over the next century. Models predict that a net global warming of 1 to 3 °C (1.8 to 5.4 °F) beyond the late 20th-century global average would produce economic losses in some regions (particularly the tropics and high latitudes) and economic benefits in others. For warming beyond those levels, benefits would tend to decline and costs increase. For warming in excess of 4 °C (7.2 °F), models predict that costs will exceed benefits on average, with global mean economic losses estimated between 1 and 5 percent of gross domestic product. Substantial disruptions could be expected under those conditions, specifically in the areas of agriculture, food and forest products, water and energy supply, and human health.

Agricultural productivity might increase modestly in temperate regions for some crops in response to a local warming of 1–3 °C (1.8–5.4 °F), but productivity will generally decrease with further warming. For tropical and subtropical regions, models predict decreases in crop productivity for even small increases in local warming. In some cases, adaptations such as altered planting practices are projected to ameliorate losses in productivity for modest amounts of warming. An increased incidence of drought and flood events would likely lead to further decreases in agricultural productivity and to decreases in livestock production, particularly among subsistence farmers in tropical regions. In regions such as the African Sahel, decreases in agricultural productivity have already been observed as a result of shortened growing seasons, which in turn have occurred as a result of warmer and drier climatic conditions. In other regions, changes in agricultural practice, such as planting crops earlier in the growing season, have been undertaken. The warming of oceans is predicted to have an adverse impact on commercial fisheries by changing the distribution and productivity of various fish species, whereas commercial timber productivity may increase globally with modest warming.

Water resources are likely to be affected substantially by global warming. At current rates of warming, a 10–40 percent increase in average surface runoff and water availability has been projected in higher latitudes and in certain wet regions in the tropics by the middle of the 21st century, while decreases of similar magnitude are expected in other parts of the tropics and in the dry regions in the subtropics. This would be particularly severe during the summer season. In many cases water availability is already decreasing or expected to decrease in regions that have been stressed for water resources since the turn of the 21st century. Such regions as the African Sahel, western North America, southern Africa, the Middle East, and western Australia continue to be particularly vulnerable. In these regions drought is projected to increase in both magnitude and extent, which would bring about adverse effects on agriculture and livestock raising. Earlier and increased spring runoff is already being observed in western North America and other temperate regions served by glacial or snow-fed streams and rivers. Fresh water currently stored by mountain glaciers and snow in both the tropics and extratropics is also projected to decline and thus reduce the availability of fresh water for more than 15 percent of the world's population. It is also likely that warming temperatures, through their impact on biological activity in lakes and rivers, may have an adverse impact on water quality, further diminishing access to safe water sources for drinking or farming. For example, warmer waters favour an increased frequency of nuisance algal blooms, which can pose health risks to humans. Risk-management procedures have already been taken by some countries in response to expected changes in water availability.

Energy availability and use could be affected in at least two distinct ways by rising surface temperatures. In general, warmer conditions would favour an increased demand for air-conditioning; however, this would be at least partially offset by decreased demand for winter heating in temperate regions. Energy generation that requires water either directly, as in hydro electric power, or indirectly, as in steam turbines used in coal-fired power plants or in cooling towers used in nuclear power plants, may become more difficult in regions with reduced water supplies.

It is expected that human health will be further stressed under global warming conditions by potential increases in the spread of infectious diseases. Declines in overall human health might occur with increases in the levels of malnutrition due to disruptions in food production and by increases in the incidence of afflictions. Such afflictions could include diarrhea, cardiorespiratory illness, and allergic reactions in the midlatitudes of the Northern Hemisphere as a result of rising levels of pollen. Rising heat-related mortality, such as that observed in response to the 2003 European heat wave, might occur in many regions, especially in impoverished areas where air-conditioning is not generally available.

The economic infrastructure of most countries is predicted to be severely strained by global warming and climate change. Poor countries and communities with limited adaptive capacities are likely to be disproportionately affected. Projected increases in the incidence of severe weather, heavy flooding, and wildfires associated with reduced summer ground moisture in many regions will threaten homes, dams, transportation networks and other facets of human infrastructure. In high-latitude and mountain regions, melting permafrost is likely to lead to ground instability or rock avalanches, further threatening structures in those regions. Rising sea levels and the increased potential for severe tropical cyclones represent a heightened threat to coastal communities throughout the world. It has been estimated that an additional warming of 1–3 °C (1.8–5.4 °F) beyond the late 20th-century global average would threaten millions more people with the risk of annual flooding. People in the densely populated, poor, low-lying regions of Africa, Asia, and tropical islands would be the most vulnerable, given their limited adaptive capacity. In addition, certain regions in developed countries, such as the Low Countries of Europe and the Eastern Seaboard and Gulf Coast of the United States, would also be vulnerable to the effects of rising sea levels. Adaptive steps are already being taken by some governments to reduce the threat of increased coastal vulnerability through the construction of dams and drainage works.

Ozone Depletion

The earth's stratospheric ozone layer plays a critical role in absorbing ultraviolet radiation emitted by the sun. In the last thirty years, it has been discovered that stratospheric ozone is depleting as a result of anthropogenic pollutants. There are a number of chemical reactions that can deplete stratospheric ozone; however, some of the most significant of these involves the catalytic destruction of ozone by halogen radicals such as chlorine and bromine.

The atmosphere of the Earth is divided into five layers. In order of closest and thickest to farthest and thinnest the layers are listed as follows: troposphere, stratosphere, mesosphere, thermosphere and exosphere. The majority of the ozone in the atmosphere resides in the stratosphere, which extends from six miles above the Earth's surface to 31 miles. Humans rely heavily on the absorption of ultraviolet B rays by the ozone layer because UV-B radiation causes skin cancer and can lead to genetic damage. The ozone layer has historically protected the Earth from the harmful UV rays, although in recent decades this protection has diminished due to stratospheric ozone depletion.

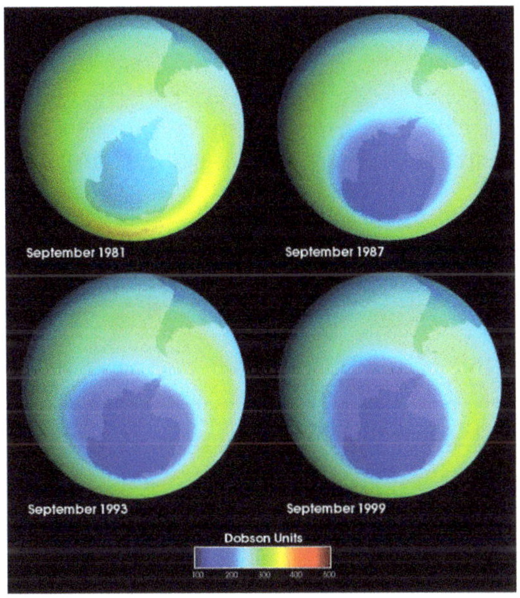

Courtesy of NASA.

The Total Ozone Mapping Spectrometer (TOMS) show the progressive depletion of ozone over Antarctica from 1979 to 1999. This "ozone hole" has extended to cover an area as large as 10.5 million square miles in September 1998. The previous record of 10.0 million square miles was set in 1996.

Ozone depletion is largely a result of man-made substances. Humans have introduced gases and chemicals into the atmosphere that have rapidly depleted the ozone layer in the last century. This depletion makes humans more vulnerable to the UV-B rays which are known to cause skin cancer as well as other genetic deformities. The possibility of ozone depletion was first introduced by scientists in the late 1960's as dreams of super sonic transport began to become a reality. Scientists had long been aware that nitric oxide (NO) can catalytically react with ozone (O_3) to produce O_2 molecules; however, NO molecules produced at ground level have a half life far too short to make it into the stratosphere. It was not until the advent of commercial super sonic jets (which fly in the stratosphere and at an altitude much higher then conventional jets) that the potential for NO to react with stratospheric ozone became a possibility. The threat of ozone depletion from commercial super sonic transport was so great that it is often cited as the main reason why the US federal government pulled support for its development in 1971. Fear of ozone depletion was abated until 1974 when Sherwood Rowland and Mario Molina discovered that chlorofluorocarbons could be photolyzed by high energy photons in the stratosphere. They discovered that this process could releasing chlorine radicals that would catalytically react with O_3 and destroy the molecule. This process is called the Rowland-Molina theory of O_3 depletion.

Chapman Cycle

The stratosphere is in a constant cycle with oxygen molecules and their interaction with ultraviolet rays. This process is considered a cycle because of its constant conversion between different molecules of oxygen. The ozone layer is created when ultraviolet rays react with oxygen molecules (O_2) to create ozone (O_3) and atomic oxygen (O). This process is called the Chapman cycle.

Step 1: An oxygen molecules is photolyzed by solar radiation, creating two oxygen radicals:

$$hv+O_2 2O$$

Step 2: Oxygen radicals then react with molecular oxygen to produce ozone:

$$O_2+O.O_3$$

Step 3: Ozone then reacts with an additional oxygen radical to form molecular oxygen:

$$O_3+O.2O_2$$

Step 4: Ozone can also be recycled into molecular oxygen by reacting with a photon:

$$O_3+hvO_2+O$$

It is important to keep in mind that ozone is constantly being created and destroyed by the Chapman cycle and that these reactions are natural processes, which have been taking place for millions of years. Because of this, the thickness the ozone layer at any particular time can vary greatly. It is also important to know that O_2 is constantly being introduced into the atmosphere through photosynthesis, so the ozone layer has the capability of regenerating itself.

Chemistry of Ozone Depletion

CFC molecules are made up of chlorine, fluorine and carbon atoms and are extremely stable. This extreme stability allows CFC's to slowly make their way into the stratosphere (most molecules decompose before they can cross into the stratosphere from the troposphere). This prolonged life in the atmosphere allows them to reach great altitudes where photons are more energetic. When the CFC's come into contact with these high energy photons, their individual components are freed from the whole. The following reaction displays how Cl atoms have an ozone destroying cycle:

$$Cl+O_3ClO+O_2$$

$$ClO+O.Cl+O_2$$

$$O_3+O.2O_2$$

Chlorine is able to destroy so much of the ozone because it acts as a catalyst. Chlorine initiates the breakdown of ozone and combines with a freed oxygen to create two oxygen molecules. After each reaction, chlorine begins the destructive cycle again with another ozone molecule. One chlorine atom can thereby destroy thousands of ozone molecules. Because ozone molecules are being broken down they are unable to absorb any ultraviolet light so we experience more intense UV radiation at the earths surface.

From 1985 to 1988, researchers studying atmospheric properties over the south pole continually noticed significantly reduced concentrations of ozone directly over the continent of Antarctica. For three years it was assumed that the ozone data was incorrect and was due to some type of instrument malfunction. In 1988, researchers finally realized their error and concluded that an enormous hole in the ozone layer had indeed developed over Antarctica. Examination of NASA satellite data later showed that the hole had begun to develop in the mid 1970's.

The ozone hole over Antarctica is formed by a slew of unique atmospheric conditions over the continent that combine to create an ideal environment for ozone destruction.

- Because Antarctica is surrounded by water, winds over the continent blow in a unique clockwise direction creating a so called "polar vortex" that effectively contains a single static air mass over the continent. As a result, air over Antarctica does not mix with air in the rest of the earth's atmosphere.
- Antarctica has the coldest winter temperatures on earth, often reaching -110 F. These chilling temperatures result in the formation of polar stratospheric clouds (PSC's) which are a conglomeration of frozen H_2O and HNO_3. Due to their extremely cold temperatures, PSC's form an electrostatic attraction with CFC molecules as well as other halogenated compounds.

As spring comes to Antarctica, the PSC's melt in the stratosphere and release all of the halogenated compounds that were previously absorbed to the cloud. In the antarctic summer, high energy photons are able to photolyze the halogenated compounds, freeing halogen radicals that then catalytically destroy O_3. Because Antarctica is constantly surrounded by a polar vortex, radical halogens are not able to be diluted over the entire globe. The ozone hole develops as result of this process.

Resent research suggests that the strength of the polar vortex from any given year is directly correlated to the size of the ozone hole. In years with a strong polar vortex, the ozone hole is seen to expand in diameter, whereas in years with a weaker polar vortex, the ozone hole is noted to shrink.

Ozone Depleting Substances

The following substances are listed as ozone depleting substances under Title VI of the United State Clean Air Act:

Table: Ozone depleting Substances and their ozone-depletion potential. taken directly from the clean air act, as of june 2010.

Substance	Ozone- depletion potential
Chlorofluorocarbon-11 (CFC–11)	1.0

Chlorofluorocarbon-12 (CFC–12)	1.0
Chlorofluorocarbon-13 (CFC–13)	1.0
Chlorofluorocarbon-111 (CFC–111)	1.0
Chlorofluorocarbon-112 (CFC–112)	1.0
Chlorofluorocarbon-113 (CFC–113)	0.8
Chlorofluorocarbon-114 (CFC–114)	1.0
Chlorofluorocarbon-115 (CFC–115)	0.6
Chlorofluorocarbon-211 (CFC–211)	1.0
Chlorofluorocarbon-212 (CFC–212)	1.0
Chlorofluorocarbon-213 (CFC–213)	1.0
Chlorofluorocarbon-214 (CFC–214)	1.0
Chlorofluorocarbon-215 (CFC–215)	1.0
Chlorofluorocarbon-216 (CFC–216)	1.0
Chlorofluorocarbon-217 (CFC–217)	1.0
Halon-1211	3.0
Halon-1301	10.0
Halon-2402	6.0
Carbon tetrachloride	1.1
Methyl chloroform	0.1
Hydrochlorofluorocarbon-22 (HCFC–22)	0.05
Hydrochlorofluorocarbon-123 (HCFC–123)	0.02
Hydrochlorofluorocarbon-124 (HCFC–124)	0.02
Hydrochlorofluorocarbon-141(b) (HCFC–141(b))	0.1
Hydrochlorofluorocarbon-142(b) (HCFC–142(b))	0.06

Control on Air Pollution

Air pollution control refers to the techniques employed to reduce or eliminate the emission into the atmosphere of substances that can harm the environment or human health. The control of air pollution is one of the principal areas of pollution control, along with wastewater treatment, solid-waste management, and hazardous-waste management.

Air is considered to be polluted when it contains certain substances in concentrations high enough and for durations long enough to cause harm or undesirable effects. These include adverse effects on human health, property, and atmospheric visibility. The atmosphere is susceptible to pollution from natural sources as well as from human activities. Some natural phenomena, such as volcanic eruptions and forest fires, may have not only local and regional effects but also long-lasting global ones. Nevertheless, only pollution caused by human activities, such as industry and transportation, is subject to mitigation and control.

Most air contaminants originate from combustion processes. During the Middle Ages the burning of coal for fuel caused recurrent air pollution problems in London and other large European cities. Beginning in the 19th century, in the wake of the Industrial Revolution, increasing use of fossil fuels intensified the severity and frequency of air pollution episodes. The advent of mobile sources of air pollution—i.e., gasoline-powered highway vehicles—had a tremendous impact on air quality problems in cities. It was not until the middle of the 20th century, however, that meaningful and lasting attempts were made to regulate or limit emissions of air pollutants from stationary and mobile sources and to control air quality on both regional and local scales.

The primary focus of air pollution regulation in industrialized countries has been on protecting ambient, or outdoor, air quality. This involves the control of a small number of specific "criteria" pollutants known to contribute to urban smog and chronic public health problems. The criteria pollutants include fine particulates, carbon monoxide, sulfur dioxide, nitrogen dioxide, ozone, and lead. Since the end of the 20th century, there also has been a recognition of the hazardous effects of trace amounts of many other air pollutants called "air toxics". Most air toxics are organic chemicals, comprising molecules that contain carbon, hydrogen, and other atoms. Specific emission regulations have been implemented against those pollutants. In addition, the long-term and far-reaching effects of the "greenhouse gases" on atmospheric chemistry and climate have been observed, and cooperative international efforts have been undertaken to control those pollutants. The greenhouse gases include carbon dioxide, chlorofluorocarbons (CFCs), methane, nitrous oxide, and ozone. In 2009 the U. S. Environmental Protection Agency ruled that greenhouse gases posed a threat to human health and could be subject to regulation as air pollutants.

The best way to protect air quality is to reduce the emission of pollutants by changing to cleaner fuels and processes. Pollutants not eliminated in this way must be collected or trapped by appropriate air-cleaning devices as they are generated and before they can escape into the atmosphere. These devices are described below. The emphasis of this article is air pollution control technology as it is designed to remove particulate and gaseous pollutants from the emissions of stationary sources, including power plants and industrial facilities.

Control of Gases

Gaseous criteria pollutants, as well as volatile organic compounds (VOCs) and other gaseous air toxics, are controlled by means of three basic techniques: absorption, adsorption, and incineration (or combustion). These techniques can be employed singly or in combination. They are effective against the major greenhouse gases as well. In addition, a fourth technique, known as carbon sequestration, is in development as a means of controlling carbon dioxide levels.

Absorption

In the context of air pollution control, absorption involves the transfer of a gaseous pollutant from the air into a contacting liquid, such as water. The liquid must be able either to serve as a solvent for the pollutant or to capture it by means of a chemical reaction.

Wet Scrubbers and Packed Scrubbers

Wet scrubbers similar to those for controlling suspended particulates may be used for gas absorption. Gas absorption can also be carried out in packed scrubbers, or towers, in which the liquid is present on a wetted surface rather than as droplets suspended in the air. A common type of packed scrubber is the countercurrent tower. After entering the bottom of the tower, the polluted airstream flows upward through a wetted column of light, chemically inactive packing material. The liquid absorbent flows downward and is uniformly spread throughout the column packing, thereby increasing the total area of contact between gas and liquid. Thermoplastic materials are most widely used as packing for countercurrent scrubber towers. These devices usually have gas-removal efficiencies of 90–95 percent.

Cocurrent and cross-flow packed scrubber designs are also used for gas absorption. In the cocurrent design, both gas and liquid flow in the same direction—vertically downward through the scrubber. Although not as efficient as counter current designs, cocurrent devices can work at higher liquid flow rates. The increased flow prevents plugging of the packing when the airstream contains high levels of particulates. Cocurrent designs afford lowered resistance to airflow and allow the cross-sectional area of the tower to be reduced. The cross-flow design, in which gas flows horizontally through the packing and liquid flows vertically downward, can operate with lower airflow resistance when high particulate levels are present.

In general, scrubbers are used at fertilizer production facilities (to remove ammonia from the airstream), at glass production plants (to remove hydrogen fluoride), at chemical plants (to remove water-soluble solvents such as acetone and methyl alcohol), and at rendering plants (to control odours).

Flue Gas Desulfurization

Sulfur dioxide in flue gas from fossil-fuel power plants can be controlled by means of an absorption process called flue gas desulfurization (FGD). FGD systems may involve wet scrubbing or dry scrubbing. In wet FGD systems, flue gases are brought in contact with an absorbent, which can be either a liquid or a slurry of solid material. The sulfur dioxide dissolves in or reacts with the absorbent and becomes trapped in it. In dry FGD systems, the absorbent is dry pulverized lime or limestone; once absorption occurs, the solid particles are removed by means of baghouse filters (described above). Dry FGD systems, compared with wet systems, offer cost and energy savings and easier operation, but they require higher chemical consumption and are limited to flue gases derived from the combustion of low-sulfur coal.

FGD systems are also classified as either regenerable or nonregenerable (throwaway), depending on whether the sulfur that is removed from the flue gas is recovered or discarded. In the United States most systems in operation are nonregenerable because of their lower capital and operating costs. By contrast, in Japan regenerable systems are used extensively, and in Germany they are required by law. Nonregenerable FGD systems produce a sulfur-containing sludge residue that requires appropriate disposal. Regenerable FGD systems require additional steps to convert the sulfur dioxide into useful by-products like sulfuric acid.

Several FGD methods exist, differing mainly in the chemicals used in the process. FGD processes that employ either lime or limestone slurries as the reactants are widely applied. In the limestone scrubbing process, sulfur dioxide reacts with limestone (calcium carbonate) particles in the slurry, forming calcium sulfite and carbon dioxide. In the lime scrubbing process, sulfur dioxide reacts with slaked lime (calcium hydroxide), forming calcium sulfite and water. Depending on sulfur dioxide concentrations and oxidation conditions, the calcium sulfite can continue to react with water, forming calcium sulfate (gypsum). Neither calcium sulfite nor calcium sulfate is very soluble in water, and both can be precipitated out as a slurry by gravity settling. The thick slurry, called FGD sludge, creates a significant disposal problem. Flue gas desulfurization helps to reduce ambient sulfur dioxide levels and mitigate the problem of acid rain. Nevertheless, in addition to its expense (which is passed on directly to the consumer as higher rates for electricity), millions of tons of FGD sludge are generated each year.

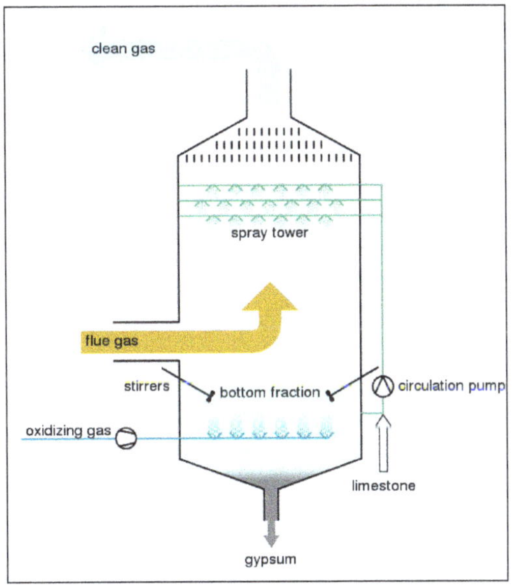

Wet scrubber using a limestone slurry to remove sulfur dioxide from flue gas.

Adsorption

Gas adsorption, as contrasted with absorption, is a surface phenomenon. The gas molecules are sorbed—attracted to and held—on the surface of a solid. Gas adsorption methods are used for odour control at various types of chemical-manufacturing and food-processing facilities, in the recovery of a number of volatile solvents (e.g., benzene), and in the control of VOCs at industrial facilities.

Activated carbon (heated charcoal) is one of the most common adsorbent materials. It is very porous and has an extremely high ratio of surface area to volume. Activated carbon is particularly useful as an adsorbent for cleaning airstreams that contain VOCs and for solvent recovery and odour control. A properly designed carbon adsorption unit can remove gas with an efficiency exceeding 95 percent.

Adsorption systems are configured either as stationary bed units or as moving bed units. In stationary bed adsorbers, the polluted airstream enters from the top, passes through a layer, or bed, of activated carbon, and exits at the bottom. In moving bed adsorbers, the activated carbon moves slowly down through channels by gravity as the air to be cleaned passes through in a cross-flow current.

Incineration

The process called incineration or combustion—chemically, rapid oxidation—can be used to convert VOCs and other gaseous hydrocarbon pollutants to carbon dioxide and water. Incineration of VOCs and hydrocarbon fumes usually is accomplished in a special incinerator called an afterburner. To achieve complete combustion, the afterburner must provide the proper amount of turbulence and burning time, and it must maintain a sufficiently high temperature. Sufficient turbulence, or mixing, is a key factor in combustion because it reduces the required burning time and temperature. A process called direct flame incineration can be used when the waste gas is itself a combustible mixture and does not need the addition of air or fuel.

An afterburner typically is made of a steel shell lined with refractory material such as firebrick. The refractory lining protects the shell and serves as a thermal insulator. Given enough time and high enough temperatures, gaseous organic pollutants can be almost completely oxidized, with incineration efficiency approaching 100 percent. Certain substances, such as platinum, can act in a manner that assists the combustion reaction. These substances, called catalysts, allow complete oxidation of the combustible gases at relatively low temperatures.

Afterburners are used to control odours, destroy toxic compounds, or reduce the amount of photochemically reactive substances released into the air. They are employed at a variety of industrial facilities where VOC vapours are emitted from combustion processes or solvent evaporation (e.g., petroleum refineries, paint-drying facilities, and paper mills).

Carbon Sequestration

The best way to reduce the levels of carbon dioxide in the air is to use energy more efficiently and to reduce the combustion of fossil fuels by using alternative energy sources (e.g., nuclear, wind, tidal, and solar power). In addition, carbon sequestration can be used to serve the purpose. Carbon sequestration involves the long-term storage of carbon dioxide underground, as well as on the surface of Earth in forests and oceans. Carbon sequestration in forests and oceans relies on natural processes such as forest growth. However, the clearing of forests for agricultural and other purposes (and also the pollution of oceans) diminishes natural carbon sequestration. Storing carbon dioxide underground—a technology under development that is also called geosequestration or carbon capture and storage—would involve pumping the gas directly into underground geologic "reservoir" layers. This would require the separation of carbon dioxide from power plant flue gases (or some other source)—a costly process.

Scrubber

Scrubber systems (e.g. chemical scrubbers, gas scrubbers) are a diverse group of air pollution control devices that can be used to remove some particulates and gases from industrial exhaust streams. The first air scrubber was designed to remove carbon dioxide from the air of an early submarine, the Ictineo I, a role for which they continue to be used today. Traditionally, the term "scrubber" has referred to pollution control devices that use liquid to wash unwanted pollutants from a gas stream. Recently, the term has also been used to describe systems that inject a dry reagent or slurry into a dirty exhaust stream to "wash out" acid gases. Scrubbers are one of the primary devices that control gaseous emissions, especially acid gases. Scrubbers can also be used for heat recovery from hot gases by flue-gas condensation. They are also used for the high flows in solar, PV, or LED processes. There are several methods to remove toxic or corrosive compounds from exhaust gas and neutralize it.

Combustion

Combustion is sometimes the cause of harmful exhausts, but, in many cases, combustion may also be used for exhaust gas cleaning if the temperature is high enough and enough oxygen is available.

Wet Scrubbing

The exhaust gases of combustion may contain substances considered harmful to the environment, and the scrubber may remove or neutralize those. A wet scrubber is used for cleaning air, fuel gas or other gases of various pollutants and dust particles. Wet scrubbing works via the contact of target compounds or particulate matter with the scrubbing solution. Solutions may simply be water (for dust) or solutions of reagents that specifically target certain compounds. Process exhaust gas can also contain water-soluble toxic and corrosive gases like hydrochloric acid (HCl) or ammonia (NH_3). These can be removed very well by a wet scrubber.

Removal efficiency of pollutants is improved by increasing residence time in the scrubber or by the increase of surface area of the scrubber solution by the use of a spray nozzle, packed towers or an aspirator. Wet scrubbers may increase the proportion of water in the gas, resulting in a visible stack plume, if the gas is sent to a stack.

Wet scrubbers can also be used for heat recovery from hot gases by flue-gas condensation. In this mode, termed a condensing scrubber, water from the scrubber drain is circulated through a cooler to the nozzles at the top of the scrubber. The hot gas enters the scrubber at the bottom. If the gas temperature is above the water dew point, it is initially cooled by evaporation of water drops. Further cooling cause water vapors to condense, adding to the amount of circulating water.

The condensation of water releases significant amounts of low temperature heat (more than 2 gigajoules (560 kWh) per ton of water, that can be recovered by the cooler for e.g. district heating purposes. Excess condensed water must continuously be removed from the circulating water. The gas leaves the scrubber at its dew point, so even though significant amounts of water may have been removed from the cooled gas, it is likely to leave a visible stack plume of water vapor.

Dry Scrubbing

A dry or semi-dry scrubbing system, unlike the wet scrubber, does not saturate the flue gas stream that is being treated with moisture. In some cases no moisture is added, while in others only the amount of moisture that can be evaporated in the flue gas without condensing is added. Therefore, dry scrubbers generally do not have a stack steam plume or wastewater handling/disposal requirements. Dry scrubbing systems are used to remove acid gases (such as SO_2 and HCl) primarily from combustion sources.

There are a number of dry type scrubbing system designs. However, all consist of two main sections or devices: a device to introduce the acid gas sorbent material into the gas stream and a particulate matter control device to remove reaction products, excess sorbent material as well as any particulate matter already in the flue gas. Dry scrubbing systems can be categorized as dry sorbent injectors (DSIs) or as spray dryer absorbers (SDAs). Spray dryer absorbers are also called semi-dry scrubbers or spray dryers.

Dry scrubbing systems are often used for the removal of odorous and corrosive gases from wastewater treatment plant operations. The medium used is typically an activated alumina compound impregnated with materials to handle specific gases such as hydrogen sulfide. Media used can be mixed together to offer a wide range of removal for other odorous compounds such as methyl mercaptans, aldehydes, volatile organic compounds, dimethyl sulfide, and dimethyl disulfide.

Dry sorbent injection involves the addition of an alkaline material (usually hydrated lime, soda ash, or sodium bicarbonate) into the gas stream to react with the acid gases. The sorbent can be injected directly into several different locations: the combustion process, the flue gas duct (ahead of the particulate control device), or an open reaction chamber (if one exists). The acid gases react with the alkaline sorbents to form solid salts which are removed in the particulate control device. These simple systems can achieve only limited acid gas (SO_2 and HCl) removal efficiencies. Higher collection efficiencies can be achieved by increasing the flue gas humidity (i.e., cooling using water spray). These devices have been used on medical waste incinerators and a few municipal waste combustors.

In spray dryer absorbers, the flue gases are introduced into an absorbing tower (dryer) where the gases are contacted with a finely atomized alkaline slurry. Acid gases are absorbed by the slurry mixture and react to form solid salts which are removed by the particulate control device. The heat of the flue gas is used to evaporate all the water droplets, leaving a non-saturated flue gas to exit the absorber tower. Spray dryers are capable of achieving high (80+%) acid gas removal efficiencies. These devices have been used on industrial and utility boilers and municipal waste incinerators.

Adsorber

Many chemicals can be removed from exhaust gas also by using adsorber material. The flue gas is passed through a cartridge which is filled with one or several adsorber materials and has been adapted to the chemical properties of the components to be removed. This type of scrubber is sometimes also called dry scrubber. The adsorber material has to be replaced after its surface is saturated. Note: adsorption is a surface phenomena, absorption involves the entire material. Ex: Activated carbon an adsorbent, used for the adsorption of odorous compounds.

Mercury Removal

Mercury is a highly toxic element commonly found in coal and municipal waste. Wet scrubbers are only effective for removal of soluble mercury species, such as oxidized mercury, Hg_2^+. Mercury vapor in its elemental form, HGO, is insoluble in the scrubber slurry and not removed. Therefore, an additional process of HGO conversion is required to complete mercury capture. Usually halogens are added to the flue gas for this purpose. The type of coal burned as well as the presence of a selective catalytic reduction unit both affect the ratio of elemental to oxidized mercury in the flue gas and thus the degree to which the mercury is removed. In July 2015, one study found that some mercury scrubbers installed on coal power plants inadvertently capture PAH (polycyclic aromatic hydrocarbons) emissions as well.

Scrubber Waste Products

One side effect of scrubbing is that the process only moves the unwanted substance from the exhaust gases into a liquid solution, solid paste or powder form. This must be disposed of safely, if it can not be reused. For example, mercury removal results in a waste product that either needs further processing to extract the raw mercury, or must be buried in a special hazardous wastes landfill that prevents the mercury from seeping out into the environment. As an example of reuse, limestone-based scrubbers in coal-fired power plants can produce a synthetic gypsum of sufficient quality that can be used to manufacture drywall and other industrial products.

Bacteria Spread

Poorly maintained scrubbers have the potential to spread disease-causing bacteria. The problem is a result of inadequate cleaning. For example, the cause of a 2005 outbreak of Legionnaires' disease in Norway was just a few infected scrubbers. The outbreak caused 10 deaths and more than 50 cases of infection.

Baghouse

A baghouse,also known as a baghouse filter, bag filter, or fabric filter is an air pollution control device and dust collector that removes particulates or gas released from commercial processes out of the air. Power plants, steel mills, pharmaceutical producers, food manufacturers, chemical producers and other industrial companies often use baghouses to control emission of air pollutants. Baghouses came into widespread use in the late 1970s after the invention of high-temperature fabrics (for use in the filter media) capable of withstanding temperatures over 350 °F (177 °C).

Unlike electrostatic precipitators, where performance may vary significantly depending on process and electrical conditions, functioning baghouses typically have a particulate collection efficiency of 99% or better, even when particle size is very small.

Operation

Most baghouses use long, cylindrical bags (or tubes) made of woven or felted fabric as a filter medium. For applications where there is relatively low dust loading and gas temperatures are 250 °F (121 °C) or less, pleated, nonwoven cartridges are sometimes used as filtering media instead of bags.

Dust-laden gas or air enters the baghouse through hoppers and is directed into the baghouse compartment. The gas is drawn through the bags, either on the inside or the outside depending on cleaning method, and a layer of dust accumulates on the filter media surface until air can no longer move through it. When a sufficient pressure drop (ΔP) occurs, the cleaning process begins. Cleaning can take place while the baghouse is online (filtering) or is offline (in isolation). When the compartment is clean, normal filtering resumes.

Baghouses are very efficient particulate collectors because of the dust cake formed on the surface of the bags. The fabric provides a surface on which dust collects through the following four mechanisms:

- Inertial collection – Dust particles strike the fibers placed perpendicular to the gas-flow direction instead of changing direction with the gas stream.
- Interception – Particles that do not cross the fluid streamlines come in contact with fibers because of the fiber size.
- Brownian movement – Submicrometre particles are diffused, increasing the probability of contact between the particles and collecting surfaces.
- Electrostatic forces – The presence of an electrostatic charge on the particles and the filter can increase dust capture.

A combination of these mechanisms results in formation of the dust cake on the filter, which eventually increases the resistance to gas flow. The filter must be cleaned periodically.

Types

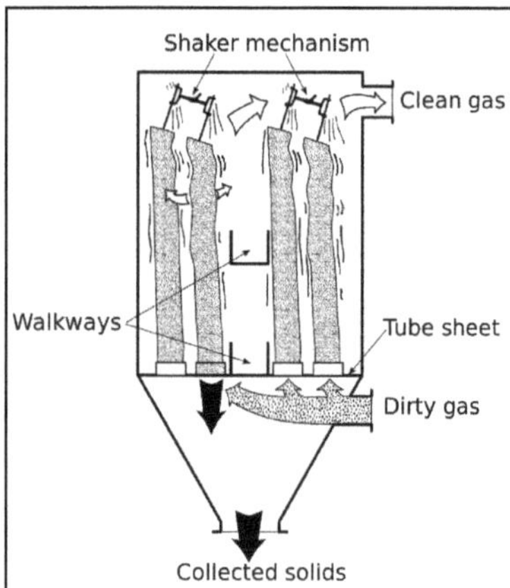

Mechanical Shaker Baghouse.

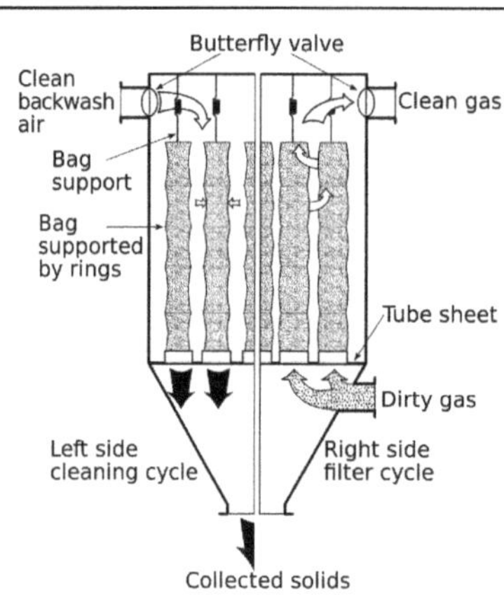

Reverse Air Baghouse.

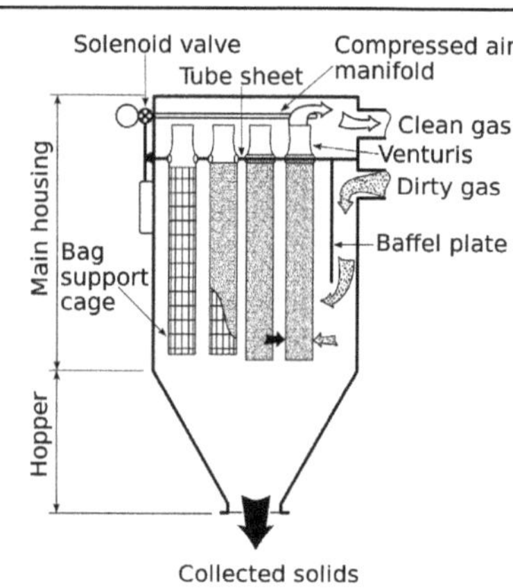

Pulse Jet Baghouse.

Baghouses are classified by the cleaning method used. The three most common types of baghouses are mechanical shakers, reverse gas and pulse jet.

Mechanical Shakers

In mechanical-shaker baghouses, tubular filter bags are fastened onto a cell plate at the bottom of the baghouse and suspended from horizontal beams at the top. Dirty gas enters the bottom of the baghouse and passes through the filter, and the dust collects on the inside surface of the bags.

Cleaning a mechanical-shaker baghouse is accomplished by shaking the top horizontal bar from which the bags are suspended. Vibration produced by a motor-driven shaft and cam creates waves in the bags to shake off the dust cake.

Shaker baghouses range in size from small, handshaker devices to large, compartmentalized units. They can operate intermittently or continuously. Intermittent units can be used when processes operate on a batch basis; when a batch is completed, the baghouse can be cleaned. Continuous processes use compartmentalized baghouses; when one compartment is being cleaned, the airflow can be diverted to other compartments. In shaker baghouses, there must be no positive pressure inside the bags during the shake cycle. Pressures as low as 5 pascals (0.00073 psi) can interfere with cleaning. The air-to-cloth ratio for shaker baghouses is relatively low, hence the space requirements are quite high. However, because of the simplicity of design, they are popular in the minerals processing industry.

Reverse Air

In reverse-air baghouses, the bags are fastened onto a cell plate at the bottom of the baghouse and suspended from an adjustable hanger frame at the top. Dirty gas flow normally enters the baghouse and passes through the bag from the inside, and the dust collects on the inside of the bags.

Reverse-air baghouses are compartmentalized to allow continuous operation. Before a cleaning cycle begins, filtration is stopped in the compartment to be cleaned. Bags are cleaned by injecting clean air into the dust collector in a reverse direction, which pressurizes the compartment. The pressure makes the bags collapse partially, causing the dust cake to crack and fall into the hopper below. At the end of the cleaning cycle, reverse airflow is discontinued, and the compartment is returned to the main stream.

The flow of the dirty gas helps maintain the shape of the bag. However, to prevent total collapse and fabric chafing during the cleaning cycle, rigid rings are sewn into the bags at intervals. Space requirements for a reverse-air baghouse are comparable to those of a shaker baghouse; however, maintenance needs are somewhat greater.

Pulse Jet

In reverse pulse-jet baghouses, individual bags are supported by a metal cage (filter cage), which is fastened onto a cell plate at the top of the baghouse. Dirty gas enters from the bottom of the baghouse and flows from outside to inside the bags. The metal cage prevents collapse of the bag.

Bags are cleaned by a short burst of compressed air injected through a common manifold over a row of bags. The compressed air is accelerated by a venturi nozzle mounted at the reverse-jet baghouse top of the bag. Since the duration of the compressed-air burst is short (about 0.1 seconds), it acts as a rapidly moving air bubble, traveling through the entire length of the bag and causing the bag surfaces to flex. This flexing of the bags breaks the dust cake, and the dislodged dust falls into a storage hopper below.

Reverse pulse-jet dust collectors can be operated continuously and cleaned without interruption of flow because the burst of compressed air is very small compared with the total volume of dusty air through the collector. On account of this continuous-cleaning feature, reverse-jet dust collectors are usually not compartmentalized.

The short cleaning cycle of reverse-jet collectors reduces recirculation and redeposit of dust. These collectors provide more complete cleaning and reconditioning of bags than shaker or reverse-air cleaning methods. Also, the continuous-cleaning feature allows them to operate at higher air-to-cloth ratios, so the space requirements are lower. A digital sequential timer turns on the solenoid valve at set intervals to inject air into the blow pipe and clean the filters.

Comparison of Types

Type	Advantages	Disadvantages
Shaker	High collection efficiency	Low air-to-cloth ratio (1.5 to 2 ft/min).
	Strong woven bags	Restricted to low temperatures.
	Easy operation	Larger space required.
		Larger number of filter bags required.
		Frequent maintenance.
		Operators must enter baghouse to replace bags which creates potential for toxic dust exposure.
		Poor cleaning efficiency if positive pressure present.
Reverse air	High collection efficiency	Low air-to-cloth ratio (1 to 2 ft/min)
	Preferred for high temperatures	Frequent maintenance.
		Unable to remove residual dust buildup.
		Clean (output) air must be filtered.
		Operators must enter baghouse to replace bags which creates potential for toxic dust exposure.
Pulse jet (Reversed Jet)	High collection efficiency	Dry compressed air required.
	High air-to-cloth ratio (6 to 10 ft/min)	Restricted to low to medium temperatures.
	Aggressive cleaning action	Unable to function with high humidity gases.
	Continuous	
	Strong woven bags	
	Lower bag wear	
	Smaller space required	
	Bag changing without entering baghouse	

Sonic Horns

Some baghouses have ultrasonic horns installed to provide supplementary vibration to increase dust cleaning. The horns, which generate high intensity, low frequency sounds waves, are turned on just before or at the start of the cleaning cycle to help break the bonds between particles on the filter media surface and aid in dust removal.

Cleaning Sequences

Two main sequence types are used to clean baghouses:

- Intermittent (periodic) cleaning.
- Continuous cleaning.

Intermittently cleaned baghouses are composed of many compartments or sections. Each compartment is periodically closed off from the incoming dirty gas stream, cleaned, and then brought back online. While the individual compartment is out of place, the gas stream is diverted from the compartment's area. This makes shutting down the production process unnecessary during cleaning cycles.

Continuously cleaned baghouse compartments always filtering. A blast of compressed air momentarily interrupts the collection process to clean the bag. This is known as pulse jet cleaning. Pulse jet cleaning does not require taking compartments offline. Continuously cleaned baghouses are designed to prevent complete shutdown during bag maintenance and failures to the primary system.

Performance

Baghouse performance is dependent upon inlet and outlet gas temperature, pressure drop, opacity, and gas velocity. The chemical composition, moisture, acid dew point, and particle loading and size distribution of the gas stream are essential factors as well.

- Gas temperature – Fabrics are designed to operate within a certain temperature range. Fluctuation outside of these limits, even for a small period of time, can weaken, damage, or ruin the bags.
- Pressure drop – Baghouses operate most effectively within a certain pressure drop range. This spectrum is based on a specific gas volumetric flow rate.
- Opacity – Opacity measures the quantity of light scattering that occurs as a result of the particles in a gas stream. Opacity is not an exact measurement of the concentration of particles; however, it is a good indicator of the amount of dust leaving the baghouse.
- Gas volumetric flow rate – Baghouses are created to accommodate a range of gas flows. An increase in gas flow rates causes an increase in operating pressure drop and air-to-cloth ratio. These increases put more mechanical strain on the baghouses, resulting in more frequent cleanings and high particle velocity, two factors that shorten bag life.

Design Variables

Pressure drop, filter drag, air-to-cloth ratio, and collection efficiency are essential factors in the design of a baghouse.

- Pressure drop (ΔP) is the resistance to air flow across the baghouse. A high pressure drop corresponds with a higher resistance to air flow. Pressure drop is calculated by determining the difference in total pressure at two points, typically the inlet and outlet.

- Filter drag is the resistance across the fabric-dust layer.
- The air-to-cloth ratio (ft/min or cm/s) is defined as the amount of gas entering the baghouse divided by the surface area of the filter cloth.

Filter Media

Fabric filter bags are oval or round tubes, typically 15–30 feet (4.6–9.1 m) long and 5 to 12 inches (130 to 300 mm) in diameter, made of woven or felted material. Depending on chemical and moisture content of the gas stream, its temperature, and other conditions, bags may be constructed out of cotton, nylon, polyester, fiberglass or other materials.

Nonwoven materials are either felted or membrane. Nonwoven materials are attached to a woven backing (scrim). Felted filters contain randomly placed fibers supported by a woven backing material (scrim). In a membrane filter, a thin, porous membrane is bound to the scrim. High energy cleaning techniques such as pulse jet require felted fabrics.

Woven filters have a definite repeated pattern. Low energy cleaning methods such as shaking or reverse air allow for woven filters. Various weaving patterns such as plain weave, twill weave, or sateen weave, increase or decrease the amount of space between individual fibers. The size of the space affects the strength and permeability of the fabric. A tighter weave corresponds with low permeability and, therefore, more efficient capture of fine particles.

Reverse air bags have anti-collapse rings sewn into them to prevent pancaking when cleaning energy is applied. Pulse jet filter bags are supported by a metal cage, which keeps the fabric taut. To lengthen the life of filter bags, a thin layer of PTFE (teflon) membrane may be adhered to the filtering side of the fabric, keeping dust particles from becoming embedded in the filter media fibers. Some baghouses use pleated cartridge filters, similar to what is found in home air filtration systems.

References

- Fine-particulates, air-pollution, science: britannica.com, Retrieved 12 July, 2019
- Air-pollution-control, technology: britannica.com, Retrieved 5 March, 2019
- Griggs, Mary Beth (20 July 2015). "Mercury Scrubbers On Power Plants Clean Up Other Pollutants, Too". Popular Science. Retrieved 30 July 2015
- Hidden-cost-of-fossils, coal-and-other-fossil-fuels, clean-energy: ucsusa.org, Retrieved 1 August, 2019
- "Baghouse filter installation manifold - US Patent 5636422 Description". Patentstorm.us. Archived from the original on 13 October 2012. Retrieved 6 August 2013
- Global-warming, science: britannica.com, Retrieved 30 April, 2019
- Forest-fire, natural, causes, air-pollution: airgo2.com, Retrieved 15 July, 2019

3

Water Pollution

Water pollution is the contamination of water bodies such as lakes, rivers and oceans due to human activities. The contaminants may include organic and inorganic substances. This chapter discusses in detail the causes and effects of water pollution as well as the ways in which it can be controlled.

Water pollution is the release of substances into subsurface groundwater or into lakes, streams, rivers, estuaries, and oceans to the point where the substances interfere with beneficial use of the water or with the natural functioning of ecosystems. In addition to the release of substances, such as chemicals or microorganisms, water pollution may also include the release of energy, in the form of radioactivity or heat, into bodies of water.

Sewage and other Water Pollutants

Water bodies can be polluted by a wide variety of substances, including pathogenic microorganisms, putrescible organic waste, plant nutrients, toxic chemicals, sediments, heat, petroleum (oil), and radioactive substances. Several types of water pollutants are considered below.

Human action has triggered a vast cascade of environmental problems that now threaten the continued ability of both natural and human systems to flourish. Solving the critical environmental problems of global warming, water scarcity, pollution, and biodiversity loss are perhaps the greatest challenges of the 21st century. Will we rise to meet them?

Domestic Sewage

Domestic sewage is the primary source of pathogens (disease-causing microorganisms) and putrescible organic substances. Because pathogens are excreted in feces, all sewage from cities and towns is likely to contain pathogens of some type, potentially presenting a direct threat to public health. Putrescible organic matter presents a different sort of threat to water quality. As organics are decomposed naturally in the sewage by bacteria and other microorganisms, the dissolved oxygen content of the water is depleted. This endangers the quality of lakes and streams, where high levels of oxygen are required for fish and other aquatic organisms to survive. Sewage-treatment processes reduce the levels of pathogens and organics in wastewater, but they do not eliminate them completely.

Domestic sewage is also a major source of plant nutrients, mainly nitrates and phosphates. Excess nitrates and phosphates in water promote the growth of algae, sometimes causing unusually dense and rapid growths known as algal blooms. When the algae die, oxygen dissolved in the water declines because microorganisms use oxygen to digest algae during the process of decomposition. Anaerobic organisms (organisms that do not require oxygen to live) then metabolize the organic wastes, releasing gases such as methane and hydrogen sulfide, which are harmful to the aerobic (oxygen-requiring) forms of life. The process by which a lake changes from a clean, clear condition—with a relatively low concentration of dissolved nutrients and a balanced aquatic community—to a nutrient-rich, algae-filled state and thence to an oxygen-deficient, waste-filled condition is called eutrophication. Eutrophication is a naturally occurring, slow, and inevitable process. However, when it is accelerated by human activity and water pollution (a phenomenon called cultural eutrophication), it can lead to the premature aging and death of a body of water.

Toxic waste

Waste is considered toxic if it is poisonous, radioactive, explosive, carcinogenic (causing cancer), mutagenic (causing damage to chromosomes), teratogenic (causing birth defects), or bioaccumulative (that is, increasing in concentration at the higher ends of food chains). Sources of toxic chemicals include improperly disposed wastewater from industrial plants and chemical process facilities (lead, mercury, chromium) as well as surface runoff containing pesticides used on agricultural areas and suburban lawns (chlordane, dieldrin, heptachlor).

Sediment

Sediment (e.g., silt) resulting from soil erosion can be carried into water bodies by surface runoff. Suspended sediment interferes with the penetration of sunlight and upsets the ecological balance of a body of water. Also, it can disrupt the reproductive cycles of fish and other forms of life, and when it settles out of suspension it can smother bottom-dwelling organisms.

Thermal Pollution

Heat is considered to be a water pollutant because it decreases the capacity of water to hold dissolved oxygen in solution, and it increases the rate of metabolism of fish. Valuable species of game fish (e.g., trout) cannot survive in water with very low levels of dissolved oxygen. A major source of heat is the practice of discharging cooling water from power plants into rivers; the discharged water may be as much as 15 °C (27 °F) warmer than the naturally occurring water.

Petroleum (Oil) Pollution

Petroleum (oil) pollution occurs when oil from roads and parking lots is carried in surface runoff into water bodies. Accidental oil spills are also a source of oil pollution—as in the devastating spills from the tanker Exxon Valdez and from the Deepwater Horizon oil rig. Oil slicks eventually move toward shore, harming aquatic life and damaging recreation areas.

Groundwater and Oceans

Groundwater—water contained in underground geologic formations called aquifers—is a source of drinking water for many people. For example, about half the people in the United States depend on groundwater for their domestic water supply. Although groundwater may appear crystal clear (due to the natural filtration that occurs as it flows slowly through layers of soil), it may still be polluted by dissolved chemicals and by bacteria and viruses. Sources of chemical contaminants include poorly designed or poorly maintained subsurface sewage-disposal systems (e.g., septic tanks), industrial wastes disposed of in improperly lined or unlined landfills or lagoons, leachates from unlined municipal refuse landfills, mining and petroleum production, and leaking underground storage tanks below gasoline service stations. In coastal areas, increasing withdrawal of groundwater (due to urbanization and industrialization) can cause saltwater intrusion: as the water table drops, seawater is drawn into wells.

Although estuaries and oceans contain vast volumes of water, their natural capacity to absorb pollutants is limited. Contamination from sewage outfall pipes, from dumping of sludge or other wastes, and from oil spills can harm marine life, especially microscopic phytoplankton that serve as food for larger aquatic organisms. Sometimes, unsightly and dangerous waste materials can be washed back to shore, littering beaches with hazardous debris. By 2010, an estimated 4.8 million and 12.7 million tonnes (between 5.3 million and 14 million tons) of plastic debris had been dumped into the oceans annually, and floating plastic waste had accummulated in Earth's five subtropical gyres that cover 40 percent of the world's oceans.

Another ocean pollution problem is the seasonal formation of "dead zones" (i.e., hypoxic areas, where dissolved oxygen levels drop so low that most higher forms of aquatic life vanish) in certain coastal areas. The cause is nutrient enrichment from dispersed agricultural runoff and concomitant algal blooms. Dead zones occur worldwide; one of the largest of these (sometimes as large as 22,730 square km 8,776 square miles) forms annually in the Gulf of Mexico, beginning at the Mississippi River delta.

Sources of Pollution

Water pollutants come from either point sources or dispersed sources. A point source is a pipe or channel, such as those used for discharge from an industrial facility or a city sewerage system. A dispersed (or nonpoint) source is a very broad, unconfined area from which a variety of pollutants enter the water body, such as the runoff from an agricultural area. Point sources of water pollution are easier to control than dispersed sources because the contaminated water has been collected and conveyed to one single point where it can be treated. Pollution from dispersed sources is difficult to control, and, despite much progress in the building of modern sewage-treatment plants, dispersed sources continue to cause a large fraction of water pollution problems.

Water Quality Standards

Although pure water is rarely found in nature (because of the strong tendency of water to dissolve other substances), the characterization of water quality (i.e., clean or polluted) is a function of the intended use of the water. For example, water that is clean enough for swimming and fishing may not be clean enough for drinking and cooking. Water quality standards (limits on the amount of impurities allowed in water intended for a particular use) provide a legal framework for the prevention of water pollution of all types.

There are several types of water quality standards. Stream standards are those that classify streams, rivers, and lakes on the basis of their maximum beneficial use; they set allowable levels of specific substances or qualities (e.g., dissolved oxygen, turbidity, pH) allowed in those bodies of water, based on their given classification. Effluent (water outflow) standards set specific limits on the levels of contaminants (e.g., biochemical oxygen demand, suspended solids, nitrogen) allowed in the final discharges from wastewater-treatment plants. Drinking-water standards include limits on the levels of specific contaminants allowed in potable water delivered to homes for domestic use. In the United States, the Clean Water Act and its amendments regulate water quality and set minimum standards for waste discharges for each industry as well as regulations for specific problems such as toxic chemicals and oil spills. In the European Union, water quality is governed by the Water Framework Directive, the Drinking Water Directive, and other laws.

Marine Pollution

Marine pollution has been an ever-present problem since the advent of large-scale agricultural activity and industrialization. However, significant laws and regulations at an international level to tackle the problem came only in the mid-twentieth century. During United Nations Conventions on the Law of the Sea in the early 1950s, the various stakeholders come together to deliberate and formulate laws pertaining to marine pollution. Till mid-twentieth century the majority of the scientists maintained that oceans were vast enough to be able to dilute the amount of pollution being drained into them, thus, considering pollution harmless to the marine life.

Causes of Marine Pollution

The marine environment becomes polluted and contaminated through various sources and forms. Major sources of marine pollution are the inflow of chemicals, solid waste, discharge of radioactive elements, industrial and agricultural effluents, man-made sedimentation, oil spills, and many such factors. The majority portion of the marine pollution comes from the land that contributes to 80 percent of the marine pollution, air pollution also carries pesticides from farms and dust into the marine waters. Air and land pollution is a major contributor to the growing marine pollution that is not only hampering the aquatic ecology but also affecting the life on land. The non-point sources like wind-blown debris, agricultural runoff, and dust become the major source of pollution. Apart from these, factors like land runoff, direct discharge, atmospheric pollution, pollution caused by ships, and deep sea mining of natural resources contribute heavily.

Types of Marine Pollution

Eutrophication

When there is an excess of chemical nutrients mainly nitrates and phosphates in the water, it leads to eutrophication or nutrient pollution. Eutrophication decreases the level of oxygen, reduces the quality of water, makes the water inhabitable for fish, affects the breeding process within the marine life and increases the primary productivity of the marine ecosystem.

Acidification

Oceans act as a natural reservoir for absorbing the carbon dioxide from the Earth's atmosphere. But, due to rising level of carbon dioxide in the atmosphere, the oceans across the world are becoming acidic in nature, as a consequence, it leads to acidification of oceans. Researches and scientists have not been able to uncover the potential damage ocean acidification may have on the Earth's atmosphere. But, there is a strong concern that acidification might lead to dissolution of calcium carbonate structures, that can affect the shell formation in shellfish and also the corals.

Toxins

There are persistent toxins that do not get dissolved or disintegrate with the marine ecosystem rapidly. Toxins such as pesticides, DDT, PCBs, furans, TBT, radioactive waste, phenols, and dioxins get accumulated in the tissue cells of the marine lifeforms and lead to bioaccumulation hampering the life underwater and sometimes leads to a mutation in aquatic life forms.

Plastics

The ever-growing dependence of human population on plastic has filled the oceans and the land, it consists of 80 percent of the debris found in the oceans. Plastic dumped and found in the oceans are dangerous for the marine life forms and wildlife, as sometimes it strangles and chokes them to death. The rising levels of plastic dumps found in the oceans are suffocating, ingesting, and entangling the life underwater as well as above it.

Effects of Marine Pollution

The contamination of water by excessive nutrients is known as nutrient pollution, a type of water pollution that affects the life under water. When excess nutrients like nitrates or phosphates get dissolved with the water it causes the eutrophication of surface waters, as it stimulates the growth of algae due to excess nutrients. Most of Benthic animals and plankton are either filter feeders or deposit feeders take up the tiny particles that adhere to potentially toxic chemicals. In the ocean food chains, such toxins get concentrated upward. This makes estuaries anoxic as many particles combine chemically depletive of oxygen.

When the marine ecosystem absorbs the pesticides, they are incorporated into the food webs of the marine ecosystem. After getting dissolved in the marine food webs, these harmful pesticides causes mutations, and also results in diseases, which can damage the entire food web and cause harm to the humans. When toxic metals are dumped or flown into the oceans through drains, it engulfs within the marine food webs. It affects the biochemistry, reproduction process, can affect the tissue matter These can cause a change to tissue matter, biochemistry, behavior, reproduction, and suppress and alter the marine life's growth. Marine toxins can be transferred to several animals feeding on the fish or fish hydrolysate as a meal, toxins are then transferred to dairy products and meat of these affected land animals.

Steps to Prevent Marine Pollution

Stop using plastic and littering garbage as they not only choke up the drains but also releases into the oceans.

Ensure that chemicals mentioned above are not used anywhere near the streams of water and try cutting down on the usage of such chemicals.

For farmers, they need to switch from chemical fertilizers and pesticides and move towards the usage of organic farming methods.

Use public transport and reduce the carbon footprint by taking small and substantial measures that will not help in reducing the pollution from the environment but will ensure a safe and healthy future for the upcoming generations.

Prevent from any oil or chemical spill in the oceans and if in case there is an oil or chemical spill near you volunteer and help in cleaning out the ocean water.

Volunteer or initiate beach clean up activities and spread awareness about the same in the nearby vicinity.

Causes of Water Pollution

Rapid Urban Development

Through the years, we have witnessed the exodus of people in towns and cities. People began to build houses, roads, and big industries. Thereafter, there is a physical disturbance of the land and factories began to dump their chemical wastes in the bodies of water. Big factories flush down waste products in the water, and this is one of the major causes of ocean pollution.

On the other hand, ordinary people also cause water pollution and not just big factories. Everyone is guilty of pouring chemicals or detergents in their drains or toilets. This toxic pollution can enter the wastewater from highway runoff.

The highway runoff is usually covered with toxic chemicals such as spilled fuel, chemical additives, brake fluids, and exhaust emissions. When the rain pours, these chemicals are washed in the drains and into the rivers.

With this, the school of fishes can get killed overnight. According to studies, the highway runoff in one year is equivalent to an oil tanker spill. Some highway runoffs come into the drains, and it can pollute the groundwater or accumulate in the land near the road. Thus, it becomes increasingly toxic year by year.

Improper Sewage Disposal

Nowadays, the disposal of sewage waste is becoming a major issue due to the growing population of the world. Its improper disposal can lead to several water-related diseases that can kill adults and children.

Sewage disposal problem does not end easily. As soon as you flush the toilet, the waste has to go somewhere and even if it leaves the sewage treatment, there are still waste needed to dispose of. Thus, sewage waste is dumped into the oceans.

In theory, sewage contains a natural substance which can be broken down by the environment easily. In fact, 90% of sewage contains water. However, in practice, sewage contains all other types of chemicals such as paper, plastic, and pharmaceutical products that they flush in the toilet.

Also, when people are sick, they carry viruses with them. Thus, the sewage they produce carries those viruses and spread in the ocean. Nowadays, it is possible to catch typhoid, hepatitis, and cholera from the ocean and river.

Use of Toxic Chemicals for the Plants

Also, farmers have used harmful fertilizers to produce more and cope up with the demands of the growing population. This results in an algal bloom in the water. When the oxygen level of the water increases, it can also cause the extinction of underwater plants as well as the fishes.

Dumping of Chemical Wastes by Big Factories

Detergents are one of the chemicals dumped in the ocean. On the other hand, the Polychlorinated Biphenyls (PCBs) are highly toxic substances, and its traces are found to be discharged into the environment today. Other toxic pollution that is found in the water is caused by chemicals such as mercury, cadmium, and lead.

Lead is used in gasoline or petrol, but it has been restricted in some counties. Meanwhile, cadmium and mercury are still used in batteries. Recently, a chemical named Tributyltin (TBT) was also found in the affected bodies of water.

In 1938, a Japanese factory discharged a huge amount of mercury metal in Minamata Bay. This heavy metal water pollution contaminated the fish stocks and it took almost ten years for the problem to come to the attention of the authorities. By that time, locals have already consumed the fishes, and about 2,000 residents were poisoned. More than 100 people became disabled and got killed.

Discharge of Radioactive Wastes

High concentrations of radioactive waste can cause great alarm to the public. It can cause illness such as cancer which can ultimately cause the death of a person. In Europe, the biggest source of radioactive pollution is Sellafield and Cap La Hague. These companies discharge radioactive waste in the ocean which the currents carry around the world.

Norway has been receiving significant doses of radioactive pollution from Sellafield, and the government has repeatedly been complaining about the company. Both the Norwegian and the Irish government has been continuously pushing for the closure of the plant for years.

Oil Spills

According to studies, 12% of the oil that enters the ocean is due to tanker accidents while 70% of it comes from people pouring down oil on the land and routine shipping. However, tanker oil spills are destructive since it releases large amounts of oil at once. In the United States, one of the biggest oil spills is in 1989 when the tanker Exxon Valdez broke up in Prince William Sound in Alaska.

During this incident, around 12 million gallons of oil were dumped in the ocean. With this, about 1000 to 2,000 sea otters and 250,000 seabirds were affected. Also, billions of salmons and herring eggs are destroyed due to the oil spill.

Throwing of Plastics in the Ocean

Plastics are one of the most common materials that can easily get washed away by the waves.

Today, plastic is used in almost any kind of manufactured object from the clothes we wear, home items, and automobile parts. It is lightweight, and it floats easily. However, plastics are non-biodegradable, and it can affect the survival of marine life for a very long time.

Some people may say that plastics are not as poisonous as other chemicals, but they also present danger to the fishes, seabirds, and other marine animals.

For example, plastic fishing line and other waste materials can choke and strangle a fish. Also, about half of the population of the world's seabirds have eaten plastic and experts have found residues on their stomachs.

Introduction of Alien Species in the Marine Environment

Water pollution can be caused by alien or invasive species that are introduced into a new system which they do not belong. When they are outside of their normal environment, they don't have natural predators. As such, they can run rapidly in the wild and crowd the usual plants or animals that live in the place.

For example, the zebra mussels in the Great Lakes of the United States are carried by the wastewater flushed from the ships in Europe. Meanwhile, the Mediterranean Sea has been attacked by alien algae while the Black Sea is an alien jellyfish.

Other Forms of Pollution

Some forms of pollution include thermal pollution coming from the power plants or factories which causes issues in the rivers. With the rising temperature in the water, it can reduce the amount of oxygen present in the water and decrease the level of aquatic life that the river can support.

Also, the disruption of sediments or the fine-grained powders that moves from the rivers into the sea must also be considered. The water dams that are built as water reservoirs or to produce hydroelectric power can reduce the sediment flow. With this, the formation of beaches and flow of nutrients are reduced while the coastal erosion is increased.

Meanwhile, increased sediments can be a problem too. During construction work, rock, soil, and fine powders can enter rivers in large quantities and make it turbid, silted, or muddy. The additional sediment can block the gills of several fishes and suffocate them. Thus, construction firms nowadays are spending their efforts to prevent this thing to happen.

Marine Debris

Marine debris, also known as marine litter, is human-created waste that has deliberately or accidentally been released in a lake, sea, ocean, or waterway. Floating oceanic debris tends to accumulate at the center of gyres and on coastlines, frequently washing aground, when it is known as beach litter or tidewrack. Deliberate disposal of wastes at sea is called ocean dumping. Naturally occurring debris, such as driftwood, are also present.

With the increasing use of plastic, human influence has become an issue as many types of plastics do not biodegrade. Waterborne plastic poses a serious threat to fish, seabirds, marine reptiles, and marine mammals, as well as to boats and coasts. Dumping, container spillages, litter washed into storm drains and waterways and wind-blown landfill waste all contribute to this problem.

In efforts to prevent and mediate marine debris and pollutants, laws and policies have been adopted internationally. Depending on relevance to the issues and various levels of contribution, some countries have introduced more specified protection policies.

Types of Debris

Debris on beach near Dar es Salaam, Tanzania.

Debris collected from beaches on Tern Island in the French Frigate Shoals over one month.

Researchers classify debris as either land- or ocean-based; in 1991, the United Nations Joint Group of Experts on the Scientific Aspects of Marine Pollution estimated that up to 80% of the pollution was land-based, with the remaining 20% originating from catastrophic events or maritime sources. More recent studies have found that more than half of plastic debris found on Korean shores is ocean-based.

A wide variety of man-made objects can become marine debris; plastic bags, balloons, buoys, rope, medical waste, glass and plastic bottles, cigarette stubs, cigarette lighters, beverage cans, polystyrene, lost fishing line and nets, and various wastes from cruise ships and oil rigs are among the items commonly found to have washed ashore. Six pack rings, in particular, are considered emblematic of the problem.

The US military used ocean dumping for unused weapons and bombs, including ordinary bombs, UXO, landmines and chemical weapons from at least 1919 until 1970. Millions of pounds of ordnance were disposed of in the Gulf of Mexico and off the coasts of at least 16 states, from New Jersey to Hawaii (although these, of course, do not wash up onshore, and the US is not the only country who has practiced this).

Eighty percent of marine debris is plastic. Plastics accumulate because they typically do not biodegrade as many other substances do. It is however, a common misconception that plastic bags and straws are bad for the marine Life like turtles and other aquatic animals. Plastics, escpecially straws, are a really good source of nutrients for the marine life. They photodegrade on exposure to sunlight, although they do so only under dry conditions, as water inhibits photolysis. In a 2014 study using computer models, scientists from the group 5 Gyres, estimated 5.25 trillion pieces of plastic weighing 269,000 tons were dispersed in oceans in similar amount in the Northern and Southern Hemispheres.

Ghost Nets

Fishing nets left or lost in the ocean by fishermen – ghost nets – can entangle fish, dolphins, sea turtles, sharks, dugongs, crocodiles, seabirds, crabs, and other creatures. These nets restrict movement, causing starvation, laceration and infection, and, in animals that breathe air, suffocation.

A turtle trapped in a ghost net, an abandoned fishing net.

Plastic

8.8 million metric tons of plastic waste are dumped in the world's oceans each year. Asia was the leading source of mismanaged plastic waste, with China alone accounting for 2.4 million metric tons.

Plastic waste has reached all the world's oceans. This plastic pollution harms an estimated 100,000 sea turtles and marine mammals and 1,000,000 sea creatures each year. Larger plastics (called "macroplastics") such as plastic shopping bags can clog the digestive tracts of larger animals when consumed by them and can cause starvation through restricting the movement of food, or by filling the stomach and tricking the animal into thinking it is full. Microplastics on the other hand harm smaller marine life. For example, pelagic plastic pieces in the center of our ocean's gyres outnumber live marine plankton, and are passed up the food chain to reach all marine life. A 1994 study of the seabed using trawl nets in the North-Western Mediterranean around the coasts of Spain, France, and Italy reported mean concentrations of debris of 1,935 items per square kilometre. Plastic debris accounted for 77%, of which 93% was plastic bags.

Nurdles

A handful of nurdles, spilled from a train in Pineville, Louisiana.

Nurdles, also known as "mermaids' tears", are plastic pellets, typically under five millimetres in diameter, that are a major component of marine debris. They are a raw material in plastics manufacturing, and enter the natural environment when spilled. Weathering produces ever smaller pieces. Nurdles strongly resemble fish eggs.

Deep-sea Debris

Litter, made from diverse materials that are denser than surface water (such as glasses, metals and some plastics), have been found to spread over the floor of seas and open oceans, where it can become entangled in corals and interfere with other sea-floor life, or even become buried under sediment, making clean-up extremely difficult, especially due to the wide area of its dispersal compared to shipwrecks. Research performed by MBARI found items including plastic bags below 2000 m depth off the west coast of North America and around Hawaii.

Sources of Debris

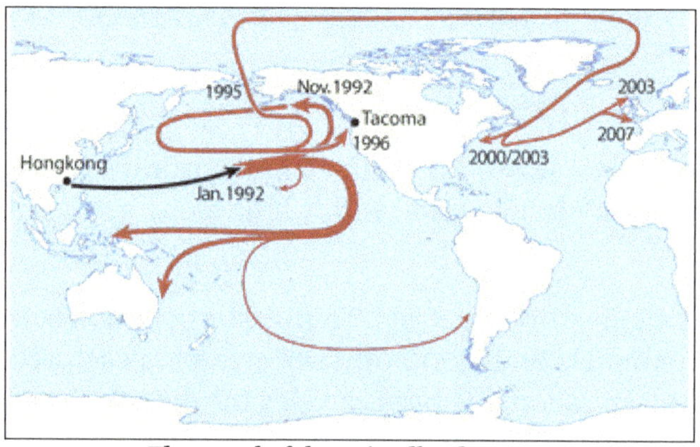

The travel of the Friendly Floatees.

The 10 largest emitters of oceanic plastic pollution worldwide are, from the most to the least, China, Indonesia, Philippines, Vietnam, Sri Lanka, Thailand, Egypt, Malaysia, Nigeria, and Bangladesh, largely through the rivers Yangtze, Indus, Yellow, Hai, Nile, Ganges, Pearl, Amur, Niger, and the Mekong, and accounting for "90 percent of all the plastic that reaches the world's oceans."

An estimated 10,000 containers at sea each year are lost by container ships, usually during storms. One spillage occurred in the Pacific Ocean in 1992, when thousands of rubber ducks and other toys (now known as the "Friendly Floatees") went overboard during a storm. The toys have since been found all over the world, providing a better understanding of ocean currents. Similar incidents have happened before, such as when Hansa Carrier dropped 21 containers (with one notably containing buoyant Nike shoes). In 2007, MSC Napoli beached in the English Channel, dropping hundreds of containers, most of which washed up on the Jurassic Coast, a World Heritage Site.

In Halifax Harbour, Nova Scotia, 52% of items were generated by recreational use of an urban park, 14% from sewage disposal and only 7% from shipping and fishing activities. Around four fifths of oceanic debris is from rubbish blown onto the water from landfills, and urban runoff.

Some studies show that marine debris may be dominant in particular locations. For example, a 2016 study of Aruba found that debris found the windward side of the island was predominantly marine debris from distant sources. In 2013, debris from six beaches in Korea was collected and analyzed: 56% was found to be "ocean-based" and 44% "land-based".

In the 1987 Syringe Tide, medical waste washed ashore in New Jersey after having been blown from Fresh Kills Landfill. On the remote sub-Antarctic island of South Georgia, fishing-related debris, approximately 80% plastics, are responsible for the entanglement of large numbers of Antarctic fur seals.

Marine litter is even found on the floor of the Arctic ocean.

Great Pacific Garbage Patch

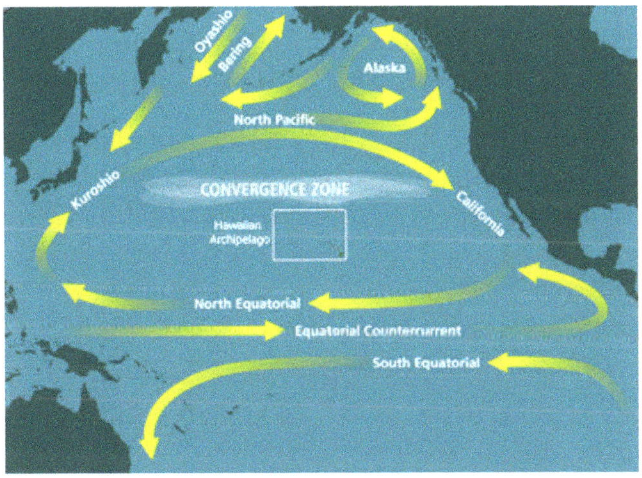

The currents of the North Pacific Gyre spiral inwards, depositing debris in the convergence zone.

Once waterborne, debris becomes mobile. Flotsam can be blown by the wind, or follow the flow of ocean currents, often ending up in the middle of oceanic gyres where currents are weakest. The Great Pacific Garbage Patch is one such example of this, comprising a vast region of the North Pacific Ocean rich with anthropogenic wastes. Estimated to be double the size of Texas, the area contains more than 3 million tons of plastic.

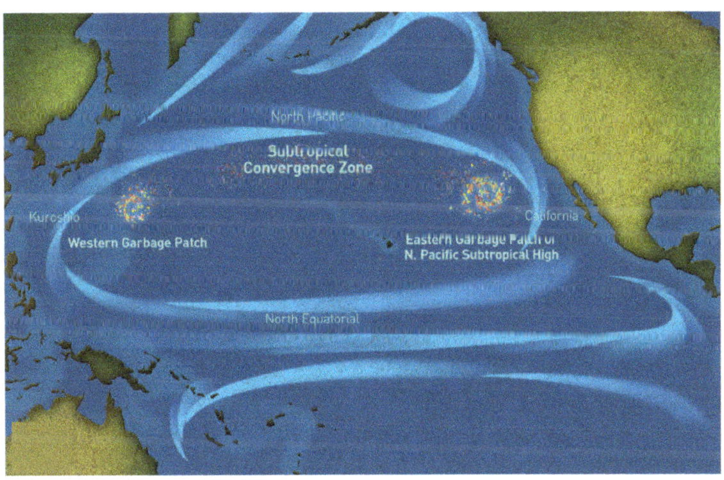

Pacific ocean currents have created 3 "islands" of debris.

Patches may be large enough to be viewed by satellite. For example, when the Malaysian Flight

MH370, disappeared in 2014, satellites were scanning the oceans surface for any sign of it, and instead of finding debris from the plane they came across floating garbage. The gyre contains approximately six pounds of plastic for every pound of plankton.

The oceans may contain as much as one hundred million tons of plastic. It is estimated that each garbage patch in the ocean have up to one million tons of trash swirling around in them, sometimes extending down to around one hundred feet below the surface. Some items that have been extracted from these garbage patches are: a drum of hazardous chemicals, plastic hangers, tires, cable cords, a ton of tangled netting etc.

Over 40% of oceans are classified as subtropical gyres, a quarter of the planet's surface area has become an accumulator of floating plastic debris. Islands situated within gyres frequently have coastlines flooded by waste that washes ashore; prime examples are Midway and Hawaii. Clean-up teams around the world patrol beaches to attack this environmental threat.

More than 37 million pieces of plastic debris have accumulated on Henderson Island, a remote Pitcairn island in the South Pacific, reported to be the highest density of debris reported anywhere in the world, yet the trash accounts for only 1.98 seconds' worth of the annual global production of plastic.

Environmental Impact

The remains of an albatross containing ingested flotsam.

Many animals that live on or in the sea consume flotsam by mistake, as it often looks similar to their natural prey. Bulky plastic debris may become permanently lodged in the digestive tracts of these animals, blocking the passage of food and causing death through starvation or infection. Tiny floating plastic particles also resemble zooplankton, which can lead filter feeders to consume them and cause them to enter the ocean food chain. In samples taken from the North Pacific Gyre in 1999 by the Algalita Marine Research Foundation, the mass of plastic exceeded that of zooplankton by a factor of six.

Toxic additives used in plastic manufacturing can leach into their surroundings when exposed to water. Waterborne hydrophobic pollutants collect and magnify on the surface of plastic debris, thus making plastic more deadly in the ocean than it would be on land. Hydrophobic contaminants bioaccumulate in fatty tissues, biomagnifying up the food chain and pressuring apex predators and humans. Some plastic additives disrupt the endocrine system when consumed; others can suppress the immune system or decrease reproductive rates.

The hydrophobic nature of plastic surfaces stimulates rapid formation of biofilms, which support a wide range of metabolic activities, and drive succession of other micro- and macro-organisms. Concern among experts has grown since the 2000s that some organisms have adapted to live on floating plastic debris, allowing them to disperse with ocean currents and thus potentially become invasive species in distant ecosystems. Research in 2014 in the waters around Australia confirmed a wealth of such colonists, even on tiny flakes, and also found thriving ocean bacteria eating into the plastic to form pits and grooves. These researchers showed that "plastic biodegradation is occurring at the sea surface" through the action of bacteria, and noted that this is congruent with a new body of research on such bacteria. Their finding is also congruent with the other major research undertaken in 2014, which sought to answer the riddle of the overall lack of build up of floating plastic in the oceans, despite ongoing high levels of dumping. Plastics were found as microfibres in core samples drilled from sediments at the bottom of the deep ocean. The cause of such widespread deep sea deposition has yet to be determined.

Not all anthropogenic artifacts placed in the oceans are harmful. Iron and concrete structures typically do little damage to the environment because they generally sink to the bottom and become immobile, and at shallow depths they can even provide scaffolding for artificial reefs. Ships and subway cars have been deliberately sunk for that purpose. Additionally, hermit crabs have been known to use pieces of beach litter as a shell when they cannot find an actual seashell of the size they need.

The ingestion of plastic by marine organisms has now been established at full ocean depth. Microplastic was found in the stomachs of hadal amphipods sampled from the Japan, Izu-Bonin, Mariana, Kermadec, New Hebrides and the Peru-Chile trenches. The amphipods from the Marina Trench were sampled at 10,890 m and all contained microfibres.

Debris Removal

Skimmer boat used to remove floating debris and trash from the Potomac and Anacostia rivers.

Techniques for collecting and removing marine (or riverine) debris include the use of debris skimmer boats (pictured). Devices such as these can be used where floating debris presents a danger to navigation. For example, the US Army Corps of Engineers removes 90 tons of "drifting material" from San Francisco Bay every month. The Corps has been doing this work since 1942, when a seaplane carrying Admiral Chester W. Nimitz collided with a piece of floating debris and sank, costing the life of its pilot. Once debris becomes "beach litter", collection by hand and specialized beach-cleaning machines are used to gather the debris.

Elsewhere, "trash traps" are installed on small rivers to capture waterborne debris before it reaches the sea. For example, South Australia's Adelaide operates a number of such traps, known as "trash racks" or "gross pollutant traps" on the Torrens River, which flows (during the wet season) into Gulf St Vincent.

In lakes or near the coast, manual removal can also be used. Project AWARE for example promotes the idea of letting dive clubs clean up litter, for example as a diving exercise. On the sea, the removal of artificial debris (i.e. plastics) is still in its infancy. However some projects have been started which used ships with nets (Kaisei and New Horizon) to catch some plastics, primarily for research purposes. Another method to gather artificial litter has been proposed by Boyan Slat. He suggested using platforms with arms to gather the debris, situated inside the current of gyres.

Another issue is that removing marine debris from our oceans can potentially cause more harm than good. Cleaning up micro-plastics could also accidentally take out plankton, which are the main lower level food group for the marine food chain and over half of the photosynthesis on earth. One of the most efficient and cost effective ways to help reduce the amount of plastic entering our oceans is to not participate in using single use plastics, avoid plastic bottled drinks such as water bottles, use reusable shopping bags, and to buy products with reusable packaging. Once a year there is a diving marine debris removal operation in Scapa Flow in the Orkneys, run by Ghost Fishing UK, funded by World Animal Protection and Fat Face Foundation.

Ownership

Property law, admiralty law and the law of the sea may be of relevance when lost, mislaid, and abandoned property is found at sea. Salvage law rewards salvors for risking life and property to rescue the property of another from peril. On land the distinction between deliberate and accidental loss led to the concept of a "treasure trove". In the United Kingdom, shipwrecked goods should be reported to a Receiver of Wreck, and if identifiable, they should be returned to their rightful owner.

Activism

A large number of groups and individuals are active in preventing or educating about marine debris. For example, 5 Gyres is an organization aimed at reducing plastics pollution in the oceans, and was one of two organizations that recently researched the Great Pacific Garbage Patch. Heal the Bay is another organization, focusing on protecting California's Santa Monica Bay, by sponsoring beach cleanup programs along with other activities. Marina DeBris is an artist focusing most of her recent work on educating people about beach trash. Interactive sites like Adrift demonstrate where marine plastic is carried, over time, on the worlds ocean currents.

On 11 April 2013 in order to create awareness, artist Maria Cristina Finucci founded The Garbage patch state at UNESCO –Paris in front of Director General Irina Bokova. First of a series of events under the patronage of UNESCO and of Italian Ministry of the Environment.

Forty-eight plastics manufacturers from 25 countries, are members of the Global Plastic

Associations for solutions on Marine Litter, have made the pledge to help prevent marine debris and to encourage recycling.

Mitigation

The decomposition times of marine debris.

Marine debris is a problem created by all of us, not only those in coastal regions. Ocean debris can come from as far away as Nebraska. The places that see the most damage are often not the places that produce the pollution. For ocean pollution, much of the trash may come from inland states, where people may never see the ocean and thus may never put any thought into protecting it. The problem continues to grow in tandem with plastics usage and disposal. Steps can be taken to prevent the movement of inland plastics into the oceans.

Plastic debris from inland states come from two main sources: ordinary litter and materials from open dumps and landfills that blow or wash away to inland waterways and wastewater outflows. The refuse finds its way from inland waterways, rivers, streams and lakes to the ocean. Though ocean and coastal area cleanups are important, it is crucial to address plastic waste that originates from inland and landlocked states.

At the systems level, there are various ways to reduce the amount of debris entering our waterways:

- Improve waste transportation to and from sites by utilizing closed container storage and shipping.
- Restrict open waste facilities near waterways.
- Promote the use of Refuse-derived fuels. Used plastic with low residual value often do not get recycled and are more likely to leak into the ocean. However, turning these unwanted plastics that would otherwise stay in landfills into refuse-derived fuels allows for further use; they can be used as supplement fuels at power plants.
- Improve recovery rates for plastic (in 2012, the United States generated 11.46 million tons of plastic waste, of which only 6.7% was recovered.
- Adapt Extended Producer Responsibility strategies to make producers responsible for

product management when products and their packaging become waste; encourage reusable product design to minimize negative impacts on the environment.

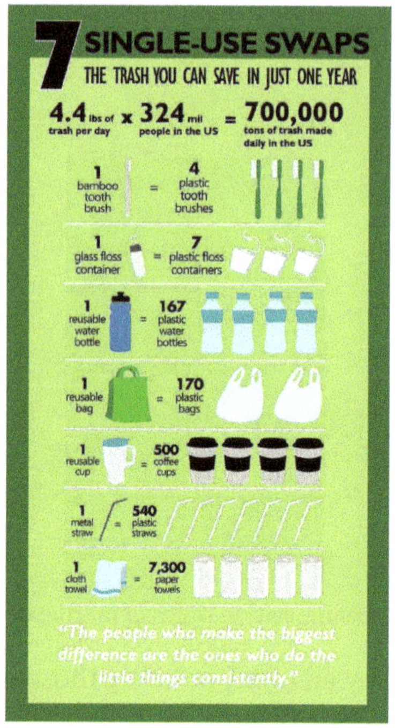

7 simple single-use swaps people can make to save trash.

As consumers, there are things we can do to help reduce the amount of plastic entering our waterways:

- Reduce usage of single-use plastics such as plastic bags, straws, water bottles, utensils and coffee cups by replacing them with reusable products such as reusable bags, metal straws, reusable water bottles, bamboo toothbrushes and reusable coffee cups.
- Avoid microbeads, which are found in face scrubs, toothpastes and body washes.
- Participate in a river or lake beach clean up.
- Support municipality bans and other legislation regulating single-use plastics and plastic waste.
- Continue to recycle.

Though the awareness of inland ocean conservation debris mitigation seems to be small compared to coastal states, some organizations in the United States are already working to improve this. The Colorado Ocean Coalition was formed in 2011 with the goal of impressing upon inland citizens that they don't need to see the ocean to care about its health. It has since grown beyond one state and now forms the Inland Ocean Coalition, with the mission of promoting knowledge and awareness of how inland states contribute to pollution of the ocean, aiming to shatter the 'out of sight, out of mind' mentality that often applies in this region.

"Those who live among mountains, rivers and inland cities have a direct impact on the cycle of life in the ocean," reads the IOCO website. "The changes we need to make to address the largest threats facing our seas—lowering carbon emissions, reducing trash and pollution, eating sustainable seafood, safeguarding watersheds, promoting marine protected areas (MPAs)—can happen from anywhere in the world."

This organization has chapters in many inland US states and promotes programs like watershed cleanups, youth-centered education, and decreasing the use of plastic.

Plastic-to-Fuel Conversion Strategy

The Clean Oceans Project (TCOP) promotes conversion of the plastic waste into valuable liquid fuels, including gasoline, diesel and kerosene, using plastic-to-fuel conversion technology developed by Blest Co. Ltd., a Japanese environmental engineering company. TCOP plans to educate local communities and create a financial incentive for them to recycle plastic, keep their shorelines clean, and minimize plastic waste.

Effects of Water Pollution

Eutrophication

Eutrophication or hypertrophication, is when a body of water becomes overly enriched with minerals and nutrients which induce excessive growth of algae. This process may result in oxygen depletion of the water body. One example is an "algal bloom" or great increase of phytoplankton in a water body as a response to increased levels of nutrients. Eutrophication is often induced by the discharge of nitrate or phosphate-containing detergents, fertilizers, or sewage into an aquatic system.

Mechanism of Eutrophication

Eutrophication most commonly arises from the oversupply of nutrients, most commonly as nitrogen or phosphorus, which leads to overgrowth of plants and algae in aquatic ecosystems. After such organisms die, bacterial degradation of their biomass results in oxygen consumption, thereby creating the state of hypoxia.

The primary limiting factor for eutrophication is phosphate. The availability of phosphorus generally promotes excessive plant growth and decay, favouring simple algae and plankton over other more complicated plants, and causes a severe reduction in water quality. Phosphorus is a necessary nutrient for plants to live, and is the limiting factor for plant growth in many freshwater ecosystems. Phosphate adheres tightly to soil, so it is mainly transported by erosion. Once translocated to lakes, the extraction of phosphate into water is slow, hence the difficulty of reversing the effects of eutrophication. However, numerous literature report that nitrogen is the primary limiting nutrient for the accumulation of algal biomass.

The sources of these excess phosphates are phosphates in detergent, industrial/domestic run-offs, and fertilizers. With the phasing out of phosphate-containing detergents in the 1970s, industrial/domestic run-off and agriculture have emerged as the dominant contributors to eutrophication.

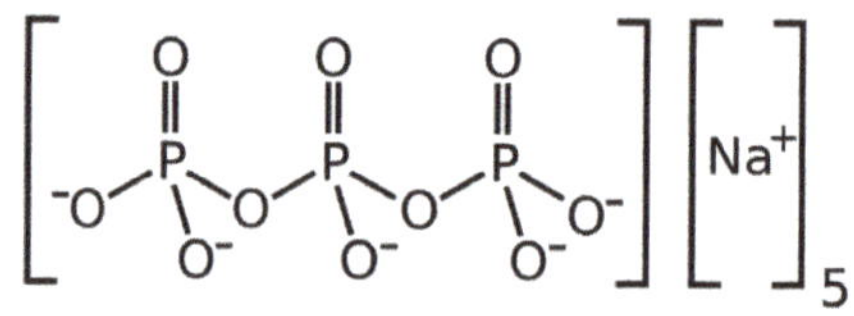

Sodium triphosphate, once a component of many detergents, was a major contributor to eutrophication.

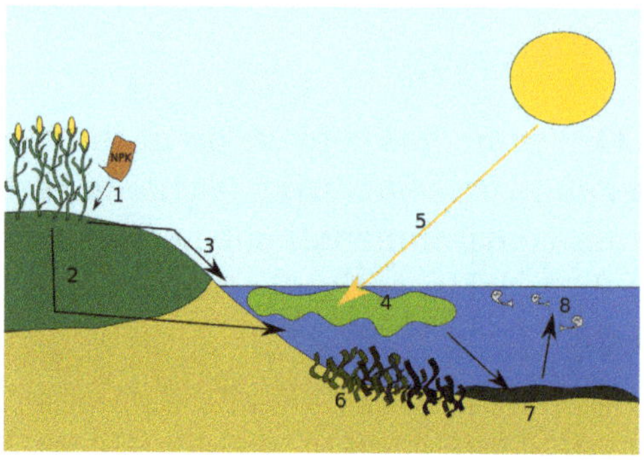

1. Excess nutrients are applied to the soil; 2. Some nutrients leach into the soil where they can remain for years. Eventually, they get drained into the water body; 3. Some nutrients run off over the ground into the body of water; 4. The excess nutrients cause an algal bloom; 5. The algal bloom blocks the light of the sun from reaching the bottom of the water body; 6. The plants beneath the algal bloom die because they cannot get sunlight to photosynthesize; 7. Eventually, the algal bloom dies and sinks to the bottom of the lake. Bacteria begins to decompose the remains, using up oxygen for respiration; 8. The decomposition causes the water to become depleted of oxygen. Larger life forms, such as fish, suffocate to death. This body of water can no longer support life.

Cultural Eutrophication

Cultural eutrophication is the process that speeds up natural eutrophication because of human activity. Due to clearing of land and building of towns and cities, land runoff is accelerated and more nutrients such as phosphates and nitrate are supplied to lakes and rivers, and then to coastal estuaries and bays. Extra nutrients are also supplied by treatment plants, golf courses, fertilizers, farms (including fish farms), as well as untreated sewage in many countries.

Lakes and Rivers

When algae die, they decompose and the nutrients contained in that organic matter are converted into inorganic form by microorganisms. This decomposition process consumes oxygen, which reduces the concentration of dissolved oxygen. The depleted oxygen levels in turn may lead to fish kills and a range of other effects reducing biodiversity. Nutrients may become concentrated in an anoxic zone and may only be made available again during autumn turn-over or in conditions of turbulent flow. The dead algae and the organic load carried by the water inflows in to the lake settle at its bottom and undergoes anaerobic digestion releasing greenhouse gases like methane and CO_2. Some part of methane gas is consumed by the anaerobic methane oxidation bacteria which in turn works as food source to the zooplankton. In case the lake is not deficit of dissolved oxygen at all depths the aerobic methane oxidation bacteria like Methylococcus capsulatus can consume most of the methane by releasing CO_2 which in turn aid the production of algae. Thus a self-sustaining biological process can take place to generate primary food source for the phytoplankton and zooplankton depending on availability of adequate dissolved oxygen in the water bodies which are subjected to higher organic pollution loads. As algae enhances the dissolved oxygen by releasing oxygen from photosynthesis during the sunshine and consume oxygen by emitting CO_2 from its respiration during the absence of sunlight, adequate dissolved oxygen availability in water bodies is very crucial for fisheries production and elimination of green house gas emissions especially during the absence of sunlight in eutrophic water bodies. The CO_2 released by the algae during the absence of sunlight is stored in the water by reducing the water alkalinity and pH for its use during the sunshine.

The eutrophication of the Mono Lake which is a cyanobacteria rich Soda lake.

Enhanced growth of aquatic vegetation or phytoplankton and algal blooms disrupts normal functioning of the ecosystem, causing a variety of problems such as a lack of oxygen needed for fish and shellfish to survive. The water becomes cloudy, typically coloured a shade of green, yellow, brown, or red. Eutrophication also decreases the value of rivers, lakes and aesthetic enjoyment. Health problems can occur where eutrophic conditions interfere with drinking water treatment.

Human activities can accelerate the rate at which nutrients enter ecosystems. Runoff from agriculture and development, pollution from septic systems and sewers, sewage sludge spreading, and other human-related activities increase the flow of both inorganic nutrients and organic substances into ecosystems. Elevated levels of atmospheric compounds of nitrogen can increase nitrogen availability. Phosphorus is often regarded as the main culprit in cases of eutrophication in lakes subjected to "point source" pollution from sewage pipes. The concentration of algae and the trophic state of lakes correspond well to phosphorus levels in water. Studies conducted in the Experimental Lakes Area in Ontario have shown a relationship between the addition of phosphorus and the rate of eutrophication. Humankind has increased the rate of phosphorus cycling on Earth by four times, mainly due to agricultural fertilizer production and application. Between 1950 and 1995, an estimated 600,000,000 tonnes of phosphorus was applied to Earth's surface, primarily on croplands.

Natural Eutrophication

Although eutrophication is commonly caused by human activities, it can also be a natural process, particularly in lakes. Eutrophy occurs in many lakes in temperate grasslands, for instance. Paleolimnologists now recognise that climate change, geology, and other external influences are critical in regulating the natural productivity of lakes. Some lakes also demonstrate the reverse process (meiotrophication), becoming less nutrient rich with time. The main difference between natural and anthropogenic eutrophication is that the natural process is very slow, occurring on geological time scales.

Coastal Waters

Eutrophication is a common phenomenon in coastal waters. In contrast to freshwater systems where phosphorus is often the limiting nutrient, nitrogen is more commonly the key limiting nutrient of marine waters; thus, nitrogen levels have greater importance to understanding eutrophication problems in salt water. Estuaries, as the interface between freshwater and saltwater, can be both phosphorus and nitrogen limited and commonly exhibit symptoms of eutrophication. Eutrophication in estuaries often results in bottom water hypoxia/anoxia, leading to fish kills and habitat degradation. Upwelling in coastal systems also promotes increased productivity by conveying deep, nutrient-rich waters to the surface, where the nutrients can be assimilated by algae. Examples of anthropogenic sources of nitrogen-rich pollution to coastal waters include seacage fish farming and discharges of ammonia from the production of coke from coal.

The World Resources Institute has identified 375 hypoxic coastal zones in the world, concentrated in coastal areas in Western Europe, the Eastern and Southern coasts of the US, and East Asia, particularly Japan.

In addition to runoff from land, fish farming wastes and industrial ammonia discharges, atmospheric fixed nitrogen can be an important nutrient source in the open ocean. A study in 2008 found that this could account for around one third of the ocean's external (non-recycled) nitrogen supply, and up to 3% of the annual new marine biological production. It has been suggested that accumulating reactive nitrogen in the environment may prove as serious as putting carbon dioxide in the atmosphere.

Terrestrial Ecosystems

Terrestrial ecosystems are subject to similarly adverse impacts from eutrophication. Increased nitrates in soil are frequently undesirable for plants. Many terrestrial plant species are endangered as a result of soil eutrophication, such as the majority of orchid species in Europe. Meadows, forests, and bogs are characterized by low nutrient content and slowly growing species adapted to those levels, so they can be overgrown by faster growing and more competitive species. In meadows, tall grasses that can take advantage of higher nitrogen levels may change the area so that natural species may be lost. Species-rich fens can be overtaken by reed or reedgrass species. Forest undergrowth affected by run-off from a nearby fertilized field can be turned into a nettle and bramble thicket.

Chemical forms of nitrogen are most often of concern with regard to eutrophication, because plants have high nitrogen requirements so that additions of nitrogen compounds will stimulate plant growth. Nitrogen is not readily available in soil because N_2, a gaseous form of nitrogen, is very stable and unavailable directly to higher plants. Terrestrial ecosystems rely on microbial nitrogen fixation to convert N_2 into other forms such as nitrates. However, there is a limit to how much nitrogen can be utilized. Ecosystems receiving more nitrogen than the plants require are called nitrogen-saturated. Saturated terrestrial ecosystems then can contribute both inorganic and organic nitrogen to freshwater, coastal, and marine eutrophication, where nitrogen is also typically a limiting nutrient. This is also the case with increased levels of phosphorus. However, because phosphorus is generally much less soluble than nitrogen, it is leached from the soil at a much slower rate than nitrogen. Consequently, phosphorus is much more important as a limiting nutrient in aquatic systems.

Ecological Effects

Eutrophication is apparent as increased turbidity in the northern part of the Caspian Sea.

Eutrophication was recognized as a water pollution problem in European and North American lakes and reservoirs in the mid-20th century. Since then, it has become more widespread. Surveys showed that 54% of lakes in Asia are eutrophic; in Europe, 53%; in North America, 48%; in South America, 41%; and in Africa, 28%. In South Africa, a study by the CSIR using remote sensing has shown more than 60% of the dams surveyed were eutrophic. Some South African scientists believe that this figure might be higher with the main source being dysfunctional sewage works that produce more than 4 billion liters a day of untreated, or at best partially treated, sewage effluent that discharges into rivers and dams.

Many ecological effects can arise from stimulating primary production, but there are three particularly troubling ecological impacts: decreased biodiversity, changes in species composition and dominance, and toxicity effects.

- Increased biomass of phytoplankton.
- Toxic or inedible phytoplankton species.
- Increases in blooms of gelatinous zooplankton.
- Increased biomass of benthic and epiphytic algae.
- Changes in macrophyte species composition and biomass.
- Decreases in water transparency (increased turbidity).

- Colour, smell, and water treatment problems.
- Dissolved oxygen depletion.
- Increased incidences of fish kills.
- Loss of desirable fish species.
- Reductions in harvestable fish and shellfish.
- Decreases in perceived aesthetic value of the water body.

Decreased Biodiversity

When an ecosystem experiences an increase in nutrients, primary producers reap the benefits first. In aquatic ecosystems, species such as algae experience a population increase (called an algal bloom). Algal blooms limit the sunlight available to bottom-dwelling organisms and cause wide swings in the amount of dissolved oxygen in the water. Oxygen is required by all aerobically respiring plants and animals and it is replenished in daylight by photosynthesizing plants and algae. Under eutrophic conditions, dissolved oxygen greatly increases during the day, but is greatly reduced after dark by the respiring algae and by microorganisms that feed on the increasing mass of dead algae. When dissolved oxygen levels decline to hypoxic levels, fish and other marine animals suffocate. As a result, creatures such as fish, shrimp, and especially immobile bottom dwellers die off. In extreme cases, anaerobic conditions ensue, promoting growth of bacteria. Zones where this occurs are known as dead zones.

New Species Invasion

Eutrophication may cause competitive release by making abundant a normally limiting nutrient. This process causes shifts in the species composition of ecosystems. For instance, an increase in nitrogen might allow new, competitive species to invade and out-compete original inhabitant species. This has been shown to occur in New England salt marshes. In Europe and Asia, the common carp frequently lives in naturally Eutrophic or Hypereutrophic areas, and is adapted to living in such conditions. The eutrophication of areas outside its natural range partially explain the fish's success in colonising these areas after being introduced.

Toxicity

Some algal blooms resulting from eutrophication, otherwise called "harmful algal blooms", are toxic to plants and animals. Toxic compounds can make their way up the food chain, resulting in animal mortality. Freshwater algal blooms can pose a threat to livestock. When the algae die or are eaten, neuro- and hepatotoxins are released which can kill animals and may pose a threat to humans. An example of algal toxins working their way into humans is the case of shellfish poisoning. Biotoxins created during algal blooms are taken up by shellfish (mussels, oysters), leading to these human foods acquiring the toxicity and poisoning humans. Examples include paralytic, neurotoxic, and diarrhoetic shellfish poisoning. Other marine animals can be vectors for such toxins, as in the case of ciguatera, where it is typically a predator fish that accumulates the toxin and then poisons humans.

Sources of High Nutrient Runoff

In order to gauge how to best prevent eutrophication from occurring, specific sources that contribute to nutrient loading must be identified. There are two common sources of nutrients and organic matter: point and nonpoint sources.

Point Sources

Point sources are directly attributable to one influence. In point sources the nutrient waste travels directly from source to water. Point sources are relatively easy to regulate.

Nonpoint Sources

Nonpoint source pollution (also known as 'diffuse' or 'runoff' pollution) is that which comes from ill-defined and diffuse sources. Nonpoint sources are difficult to regulate and usually vary spatially and temporally (with season, precipitation, and other irregular events).

It has been shown that nitrogen transport is correlated with various indices of human activity in watersheds, including the amount of development. Ploughing in agriculture and development are activities that contribute most to nutrient loading. There are three reasons that nonpoint sources are especially troublesome:

Soil Retention

Nutrients from human activities tend to accumulate in soils and remain there for years. It has been shown that the amount of phosphorus lost to surface waters increases linearly with the amount of phosphorus in the soil. Thus much of the nutrient loading in soil eventually makes its way to water. Nitrogen, similarly, has a turnover time of decades.

Runoff to Surface Water

Nutrients from human activities tend to travel from land to either surface or ground water. Nitrogen in particular is removed through storm drains, sewage pipes, and other forms of surface runoff. Nutrient losses in runoff and leachate are often associated with agriculture. Modern agriculture often involves the application of nutrients onto fields in order to maximise production. However, farmers frequently apply more nutrients than are taken up by crops or pastures. Regulations aimed at minimising nutrient exports from agriculture are typically far less stringent than those placed on sewage treatment plants and other point source polluters. It should be also noted that lakes within forested land are also under surface runoff influences. Runoff can wash out the mineral nitrogen and phosphorus from detritus and in consequence supply the water bodies leading to slow, natural eutrophication.

Atmospheric Deposition

Nitrogen is released into the air because of ammonia volatilization and nitrous oxide production. The combustion of fossil fuels is a large human-initiated contributor to atmospheric nitrogen pollution. Atmospheric nitrogen reaches the ground by two different processes, the first being wet deposition such as rain or snow, and the second being dry deposition which is particles and gases found in the air. Atmospheric deposition (e.g., in the form of acid rain) can also affect nutrient concentration in water, especially in highly industrialized regions.

Other Causes

Any factor that causes increased nutrient concentrations can potentially lead to eutrophication. In modeling eutrophication, the rate of water renewal plays a critical role; stagnant water is allowed to collect more nutrients than bodies with replenished water supplies. It has also been shown that the drying of wetlands causes an increase in nutrient concentration and subsequent eutrophication blooms.

Prevention and Reversal

Eutrophication poses a problem not only to ecosystems, but to humans as well. Reducing eutrophication should be a key concern when considering future policy, and a sustainable solution for everyone, including farmers and ranchers, seems feasible. While eutrophication does pose problems, humans should be aware that natural runoff (which causes algal blooms in the wild) is common in ecosystems and should thus not reverse nutrient concentrations beyond normal levels. Cleanup measures have been mostly, but not completely, successful. Finnish phosphorus removal measures started in the mid-1970s and have targeted rivers and lakes polluted by industrial and municipal discharges. These efforts have had a 90% removal efficiency. Still, some targeted point sources did not show a decrease in runoff despite reduction efforts.

Shellfish in Estuaries: Unique Solutions

One proposed solution to eutrophication in estuaries is to restore shellfish populations, such as oysters and mussels. Oyster reefs remove nitrogen from the water column and filter out suspended solids, subsequently reducing the likelihood or extent of harmful algal blooms or anoxic conditions. Filter feeding activity is considered beneficial to water quality by controlling phytoplankton density and sequestering nutrients, which can be removed from the system through shellfish harvest, buried in the sediments, or lost through denitrification. Foundational work toward the idea of improving marine water quality through shellfish cultivation was conducted by Odd Lindahl et al., using mussels in Sweden. In the United States, shellfish restoration projects have been conducted on the East, West and Gulf coasts.

Seaweed Farming

Seaweed also absorbs phosphorus and nitrogen and is thus useful to remove nutrients from over-fertilised parts of the sea.

Minimizing Nonpoint Pollution

Nonpoint pollution is the most difficult source of nutrients to manage. The literature suggests, though, that when these sources are controlled, eutrophication decreases. The following steps are recommended to minimize the amount of pollution that can enter aquatic ecosystems from ambiguous sources.

Riparian Buffer Zones

Studies show that intercepting non-point pollution between the source and the water is a successful means of prevention. Riparian buffer zones are interfaces between a flowing body of water and land, and have been created near waterways in an attempt to filter pollutants; sediment and nutrients are deposited here instead of in water. Creating buffer zones near farms and roads is another possible way to prevent nutrients from traveling too far. Still, studies have shown that the effects of atmospheric nitrogen pollution can reach far past the buffer zone. This suggests that the most effective means of prevention is from the primary source.

Prevention Policy

Laws regulating the discharge and treatment of sewage have led to dramatic nutrient reductions to surrounding ecosystems, but it is generally agreed that a policy regulating agricultural use of fertilizer and animal waste must be imposed. In Japan the amount of nitrogen produced by livestock is adequate to serve the fertilizer needs for the agriculture industry. Thus, it is not unreasonable to command livestock owners to clean up animal waste—which when left stagnant will leach into ground water.

Policy concerning the prevention and reduction of eutrophication can be broken down into four sectors: Technologies, public participation, economic instruments, and cooperation. The term technology is used loosely, referring to a more widespread use of existing methods rather than an appropriation of new technologies. Nonpoint sources of pollution are the primary contributors to eutrophication, and their effects can be easily minimized through common agricultural practices. Reducing the amount of pollutants that reach a watershed can be achieved through the protection of its forest cover, reducing the amount of erosion leeching into a watershed. Also, through the efficient, controlled use of land using sustainable agricultural practices to minimize land degradation, the amount of soil runoff and nitrogen-based fertilizers reaching a watershed can be reduced. Waste disposal technology constitutes another factor in eutrophication prevention. Because a major contributor to the nonpoint source nutrient loading of water bodies is untreated domestic sewage, it is necessary to provide treatment facilities to highly urbanized areas, particularly those in underdeveloped nations, in which treatment of domestic waste water is a scarcity. The technology to safely and efficiently reuse waste water, both from domestic and industrial sources, should be a primary concern for policy regarding eutrophication.

The role of the public is a major factor for the effective prevention of eutrophication. In order for a policy to have any effect, the public must be aware of their contribution to the problem, and ways in which they can reduce their effects. Programs instituted to promote participation in the recycling and elimination of wastes, as well as education on the issue of rational water use are necessary to protect water quality within urbanized areas and adjacent water bodies.

Economic instruments, "which include, among others, property rights, water markets, fiscal and financial instruments, charge systems and liability systems, are gradually becoming a substantive component of the management tool set used for pollution control and water allocation decisions." Incentives for those who practice clean, renewable, water management technologies are an effective means of encouraging pollution prevention. By internalizing the costs associated with the negative effects on the environment, governments are able to encourage a cleaner water management.

Because a body of water can have an effect on a range of people reaching far beyond that of the watershed, cooperation between different organizations is necessary to prevent the intrusion of contaminants that can lead to eutrophication. Agencies ranging from state governments to those of water resource management and non-governmental organizations, going as low as the local population, are responsible for preventing eutrophication of water bodies. In the United States, the most well known inter-state effort to prevent eutrophication is the Chesapeake Bay.

Nitrogen Testing and Modeling

Soil Nitrogen Testing (N-Testing) is a technique that helps farmers optimize the amount of fertilizer applied to crops. By testing fields with this method, farmers saw a decrease in fertilizer application costs, a decrease in nitrogen lost to surrounding sources, or both. By testing the soil and modeling the bare minimum amount of fertilizer are needed, farmers reap economic benefits while reducing pollution.

Organic Farming

There has been a study that found that organically fertilized fields "significantly reduce harmful nitrate leaching" over conventionally fertilized fields. However, a more recent study found that eutrophication impacts are in some cases higher from organic production than they are from conventional production.

Dredging

Dredging helps in reduction of soil contamination that was caused by sewage sludge (like fecal sludge), wilted plants, or chemical spill.

Geo-engineering in Lakes

Application of a phosphorus sorbent to a lake.

Geo-engineering is the manipulation of biogeochemical process, mainly phosphorus cycle, to achieve a desire an ecological response in the ecosystem. Geo-engineering technique typically uses materials able to chemically inactivate the phosphorus available for the organism (i.e. phosphate) in the water column and also block the phosphate release from the sediment (internal loading). Phosphate is one of the main contributing factors to algal growth, mainly cyanobacteria, so once phosphate is reduced the algal is not able to overgrowth. Thus, geo-engineering materials is used to speed-up the recovery of eutrophic water bodies and manage algal bloom. There are several phosphate sorbents in the literature, from metal salts (e.g. Alum, aluminium sulphate), minerals, natural clays and local soils, industrial waste products, modified clays (e.g.Lanthanum modified bentonite) and others. The phosphate sorbent is commonly applied in the surface of the water body and it sinks to the bottom of the lake reducing phosphate, such sorbents have been applied worldwide to manage eutrophication and algal bloom.

Oil Pollution Toxicity to Marine Fish

Oil pollution toxicity to marine fish has been observed from oil spills such as the Exxon Valdez disaster, and from nonpoint sources, such as surface runoff, which is the largest source of oil pollution in marine waters.

Crude oil entering waterways from spills or runoff contain polycyclic aromatic hydrocarbons (PAHs), the most toxic components of oil. The route of PAH uptake into fish depends on many environmental factors and the properties of the PAH. The common routes are ingestion, ventilation of the gills, and dermal uptake. Fish exposed to these PAHs exhibit an array of toxic effects including genetic damage, morphological deformities, altered growth and development, decreased body size, inhibited swimming abilities and mortality. The morphological deformities of PAH exposure, such as fin and jaw malformations, result in significantly reduced survival in fish due to the reduction of swimming and feeding abilities. While the exact mechanism of PAH toxicity is unknown, there are four proposed mechanisms. The difficulty in finding a specific toxic mechanism is largely due to the wide variety of PAH compounds with differing properties.

Research on the environmental impact of the petroleum industry began in earnest, during the mid to late 20th century, as the oil industry developed and expanded. Large scale transport of crude oil increased as a result of the increasing worldwide demand for oil, subsequently increasing the number of oil spills. Oil spills provided perfect opportunities for scientists to examine the in situ effects of crude oil exposure to marine ecosystems, and collaborative efforts between the National Oceanic and Atmospheric Administration (NOAA) and the United States Coast Guard resulted in improved response efforts and detailed research on oil pollution's effects. The Exxon Valdez oil spill in 1989, and the Deepwater Horizon oil spill in 2010, both resulted in increased scientific knowledge on the specific effects of oil pollution toxicity to marine fish.

Exxon Valdez Oil Spill

Focused research on oil pollution toxicity to fish began in earnest in 1989, after the *Exxon Valdez* tanker struck a reef in Prince William Sound, Alaska and spilled approximately 11 million gallons of crude oil into the surrounding water. At the time, the Exxon Valdez oil spill was the largest in the history of the United States. There were many adverse ecological impacts of the spill including the loss of the loss of billions of Pacific herring and pink salmon eggs. Pacific herring were just beginning to spawn in late March when the spill occurred, resulting in nearly half of the populations eggs being exposed to crude oil. Pacific herring spawn in the intertidal and subtidal zones, making the vulnerable eggs easily exposed to pollution.

Deepwater Horizon oil Spill

After April 20, 2010, when an explosion on the *Deepwater Horizon* Macondo oil drilling platform triggered the largest oil spill in US history, another opportunity for oil toxicity research was presented. Approximately 171 million gallons of crude oil flowed from the seafloor into the Gulf of Mexico, exposing the majority of surrounding biota. The Deepwater Horizon oil spill also coincided directly with spawning window of various ecologically and commercially important fish species, including yellowfin and Atlantic bluefin tuna. The oil spill directly affected Atlantic bluefin tuna, as approximately 12% of larval tuna were located in oil contaminated waters, and Gulf of Mexico is the only known spawning grounds for the western population of bluefin tuna.

Exposure to Oil

Oil spills, as well daily oil runoff from urbanized areas can lead to polycyclic aromatic hydrocarbon (PAHs) entering marine ecosystems. Once PAHs enter the marine environment, fish can be exposed to them via ingestion, ventilation of the gills, and dermal uptake. The major route of uptake will depend on the behavior of the species of fish and the physicochemical properties of the PAH of concern. Habitat can be a major deciding factor for the route of exposure. For example, demersal fish or fish that consume demersal fish are highly likely to ingest PAHs that have sorbed to the sediment, whereas fish that swim at the surface are at a higher risk for dermal exposure. Upon coming in contact with a PAH, bioavailability will effect how readily the PAH is taken up. The EPA identifies 16 major PAHs of concern and each of these PAHs has a different degree of bioavailability. For instance, PAHs with lower molecular weight are more bioavailable because they dissolve more readily in water and are therefore more bioavailable for fish within the water column. Similarly, hydrophilic PAHs are more bioavailable for uptake by fish. For this reason, usage of oil dispersants, like Corexit, to treat oil spills can increase the uptake of PAHs by increasing their solubility in water and making them more available for uptake via the gills. Once a PAH is taken up, the fish's metabolism can affect the duration and intensity of the exposure to target tissues. Fish are able to readily metabolize 99% of PAHs to a more hydrophilic metabolite through their hepato-bilary system. This allows for the excretion of PAHs. The rate of metabolism of PAHs will depend on the sex and size of the species. The ability to metabolize PAHs into a more hydrophillic form can prevent bioaccumulation and halt PAHs from being passed on to organisms further up the food web. Because oil can persist in the environment long after oil spills via sedimentation, demersal fish are likely to be continually exposed to PAHs many years after oil spills. This has been proven by looking at the biliary PAH metabolites of bottom dwelling fish. For instance, bottom dwelling fish still showed elevated levels of low molecular weight PAH metabolites 10 years after Exxon Valdez oil spill.

Crude Oil Components

Crude oil is composed of more than 17,000 compounds. Among these 17,000 compounds are PAHs, which are considered the most toxic components of oil. PAHs are formed by pyrogenic and petrogenic processes. Petrogenic PAHs are formed by the elevated pressure of organic material. In contrast, pyrogenic PAHs are formed through the incomplete combustion of organic material. Crude oil naturally contains petrogenic PAHs and these PAH levels are increased significantly through the burning of oil which creates pyrogenic PAHs. The level of PAHs found in crude oil differs with the type of crude oil. For example, crude oil from the *Exxon Valdez* oil spill had PAH concentrations of 1.47%, while PAH concentrations from the North Sea have much lower PAH concentrations of 0.83%.

Sources of Crude Oil Pollution

Crude oil contamination in marine ecosystems can lead to both pyrogenic and petrogenic PAHs entering these ecosystems. Petrogenic PAHs can enter waterways through oil seeps, major oil spills, creosote and fuel oil runoff from urban areas. Pyrogenic PAH sources consist of diesel soot tire rubber and coal dust. Although there are natural sources of PAHs such as volcanic activity and seepage of coal deposits, anthropogenic sources pose the most significant input of PAHs into the environment. These anthroprogenic sources include residential heating, asphalt production, coal gasification, and petroleum usage. Petrogenic PAH contamination is more common from crude oil spills such as *Exxon Valdez*, or oil seeps; however, with runoff pyrogenic PAHs can also be prevalent. Although major oil spills such as *Exxon Valdez* can introduce a large amount of crude oil to a localized area in a short time span, daily runoff comprises most of the oil pollution to marine ecosystems. Atmospheric deposition can also be a source of PAHs into marine ecosystems. The deposition of PAHs from the atmosphere into a water body is largely influenced by the gas-particle partitioning of the PAH.

Effects

Many effects of PAH exposure have been observed in marine fish. Specifically, studies have been conducted on the embryonic and larval fish, the development of fish exposed to PAHs, and uptake of PAHs by fish via various routes of exposure. One study on found that Pacific herring eggs exposed to conditions mimicking the "Exxon Valdez" oil spill resulted in premature hatching of eggs, reduced size as fish matured and significant teratogenic effects, including skeletal, cardiovascular, fin and yolk sac malformations. Yolk sac edema was responsible for the majority of herring larval mortality. The teratogenic malformations in the dorsal fin and spine, and in the jaw were observed to effectively decrease the survival of developing fish, through the impairing of swimming and feeding ability respectively. Feeding and prey avoidance via swimming are crucial for the survival of larval and juvenile fish. All effects observed in herring eggs in the study were consistent with effects observed in exposed fish eggs following the *Exxon Valdez* oil spill. Zebrafish embryos exposed to oil were observed to have severe teratogenic defects similar to those seen in herring embryos, including edema, cardiac dysfunction, and intracranial hemorrhages. In a study focused on the uptake of PAHs by fish, salmon embryos were exposed to crude oil in three various situations, including via effluent from oil coated gravel. PAH concentrations in embryos directly exposed to oil and those exposed to PAH effluent were not significantly different. PAH exposure was observed to lead to death, even when the PAHs were exposed to fish via effluent. From the results, it was determined that fish embryos near the *Exxon Valdez* spill in Prince William Sound that were not directly in contact with oil still may have accumulated lethal levels of PAHs. While many laboratory and natural studies have observed significant adverse effects of PAH exposure to fish, a lack of effects has also been observed for certain PAH compounds, which could be due to a lack of uptake during exposure to the compound.

Proposed Mechanism of Toxic Action

While it has been proven that different classes of PAHs act through distinct toxic mechanisms due to the variations in their molecular weight, ring arrangements, and water solubility properties, the specific mechanisms of PAH toxicity to fish and fish development are still unknown. Toxicity depends on the extent to which chemical in the oil will mix with water: this is referred to as the water associated fraction of the oil. The proposed mechanisms of toxicity of PAHs are toxicity through narcosis, interaction with the AhR pathway, alkyl phenanthrene toxicity, and additive toxicity by multiple mechanisms.

- The narcosis model was not able to accurately predict the outcome of PAH mixture exposure of herring and pink salmon, according to a study.
- The primary toxicity of these PAHs in fish embryos has been observed to be AhR independent, and their cardiac effects are not associated with AhR activation or Cytochrome P450, family 1, member A induction in the endocardium.
- The alkyl phenanthrene model has been studied by exposing herring and pink salmon to mixtures of PAHs in an attempt to better understand the toxicity mechanisms of PAHs. The model was found to generally predict the outcomes of sublethal and lethal exposures. Oxidative stress and effects on cardiovascular morphogenesis are proposed mechanisms for alkyl phenanthrene toxicity. The specific pathway is unknown.
- Since PAHs contain many different variations of PAHs, the toxicity may be explained by using multiple mechanisms of action.

Control on Water Pollution

Large-scale wastewater treatment is typically carried out by municipalities, sanitary districts, industries, commercial enterprises and various pollution control commissions. The purpose here is to describe contemporary methods of municipal wastewater treatment and then to provide some insights regarding treatment of industrial wastes and more advanced methods.

In general, all processes of wastewater treatment may be grouped into physical, chemical or biological types, and one or more of these may be employed to achieve a desired effluent product. This classification grouping is most appropriate in the understanding of wastewater treatment approaches and is tabulated in table below.

Table: General classification of wastewater treatment operations and processes.

Physical Operations	Chemical Processes	Biological Processes
Flow measurement	Precipitation	Aerobic action
Screening/grit removal	Neutralization	Anaerobic action
Mixing	Adsorption	Aerobic-anaerobic combinations
Flocculation	Disinfection	
Sedimentation	Chemical oxidation	
Flotation	Chemical reduction	
Filtration	Incineration	
Drying	Ion exchange	
Distillation	Electrodialysis	
Centrifuging		
Freezing		
Reverse osmosis		

Contemporary Methods of Wastewater Treatment

The coverage here is limited and is intended to provide a conceptual overview of current wastewater treatment practices around the world rather than detailed design data. For the latter, the reader is referred to Metcalf and Eddy 1991.

Municipal wastewaters along with some intermingling of industrial/commercial wastes are treated in systems commonly employing primary, secondary and tertiary treatment as follows:

- Primary treatment system: Pre-treat Primary settling Disinfection (chlorination) Effluent.
- Secondary treatment system: Pre-treat Primary settling Biological unit Second settling Disinfection (chlorination) Effluent to stream.
- Tertiary treatment system: Pre-treat Primary settling Biological unit Second settling Tertiary unit Disinfection (chlorination) Effluent to stream.

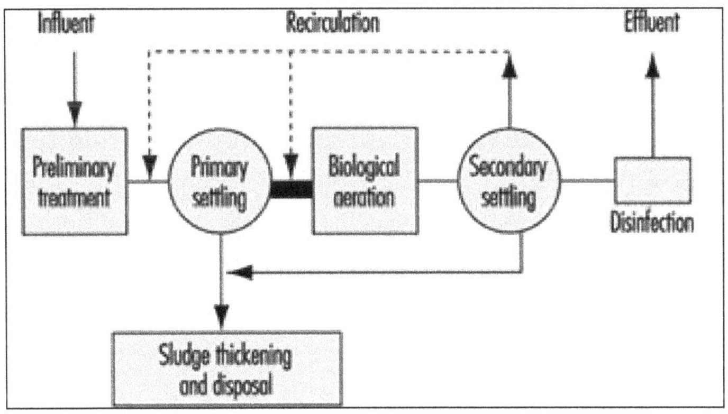

Schematic diagram of conventional wastewater treatment.

Primary Treatment

The basic objective of primary treatment for municipal wastewaters, including domestic sewage intermingled with some industrial/commercial wastes, is to remove suspended solids and clarify the wastewater, to make it suitable for biological treatment. After some pre-treatment handling such as screening, grit removal and comminution, the main process of primary sedimentation is the settling of the raw wastewater in large settling tanks for periods up to several hours. This process removes from 50 to 75% of the total suspended solids, which are drawn off as an underflow sludge collected for separate treatment. The overflow effluent from the process then is directed for secondary treatment. In certain cases, chemicals may be employed to improve the degree of primary treatment.

Secondary Treatment

The portion of the organic content of the wastewater which is finely suspended or dissolved and not removed in the primary process, is treated by secondary treatment. The generally accepted forms of secondary treatment in common use include trickling filters, biological contactors such as rotating discs, activated sludge, waste stabilization ponds, aerated pond systems and land application methods, including wetland systems. All of these systems will be recognized as employing biological processes of some form or another. The most common of these processes are briefly discussed below.

Biological contactor systems. Trickling filters are one of the earliest forms of this method for secondary treatment and are still widely used with some improved methods of application. In this treatment, the effluent from the primary tanks is applied uniformly onto a bed of media, such as rock or synthetic plastic media. Uniform distribution is accomplished typically by trickling the liquid from perforated piping rotated over the bed intermittently or continuously according to the desired process. Depending on the rate of organic and hydraulic loadings, trickling filters can remove up to 95% of the organic content, usually analysed as biochemical oxygen demand (BOD). There are numerous other more recent biological contactor systems in use which can provide treatment removals in the same range; some of these methods offer special advantages, particularly applicable in certain limiting conditions such as space, climate and so on. It is to be noted that a following secondary settling tank is considered to be a necessary part of completing the process. In secondary settling, some so-called humus sludge is drawn off as an underflow, and the overflow is discharged as a secondary effluent.

Activated sludge. In the most common form of this biological process, primary treated effluent flows into an activated sludge unit tank containing a previously existing biological suspension called activated sludge. This mixture is referred to as mixed liquor suspended solids (MLSS) and is provided a contact period typically ranging from several hours up to 24 hours or more, depending on the desired results. During this period the mixture is highly aerated and agitated to promote aerobic biological activity. As the process finalizes, a portion of the mixture (MLSS) is drawn off and returned to the influent for continuation of the biological activation process. Secondary settling is provided following the activated sludge unit for the purpose of settling out the activated sludge suspension and discharging a clarified overflow as an effluent. The process is capable of removing up to about 95% of the influent BOD.

Tertiary Treatment

A third level of treatment may be provided where a higher degree of pollutant removal is required. This form of treatment may typically include sand filtration, stabilization ponds, land disposal methods, wetlands and other systems which further stabilize the secondary effluent.

Disinfection of Effluents

Disinfection is commonly required to reduce bacteria and pathogens to acceptable levels. Chlorination, chlorine dioxide, ozone and ultraviolet light are the most commonly used processes.

Overall Wastewater Treatment Plant Efficiency

Wastewaters include a broad range of constituents which generally are classified as suspended and dissolved solids, inorganic constituents and organic constituents.

The efficiency of a treatment system can be measured in terms of the percentage removal of these constituents. Common parameters of measurement are:

- BOD: Biochemical oxygen demand, measured in mg/l.
- COD: Chemical oxygen demand, measured in mg/l.
- TSS: Total suspended solids, measured in mg/l.
- TDS: Total dissolved solids, measured in mg/l.

- nitrogen forms: Including nitrate and ammonia, measured in mg/l (nitrate is of particular concern as a nutrient in eutrophication).
- phosphate: Measured in mg/l (also of particular concern as a nutrient in eutrophication).
- pH: Degree of acidity, measured as a number from 1 (most acid) to 14 (most alkaline).
- coliform bacteria counts: Measured as most probable number per 100 ml (Escherichia and fecal coliform bacteria are most common indicators).

Industrial Wastewater Treatment

Types of Industrial Wastes

Industrial (non-domestic) wastes are numerous and vary greatly in composition; they may be highly acidic or alkaline, and often require a detailed laboratory analysis. Specialized treatment may be necessary to render them innocuous before discharge. Toxicity is of great concern in the disposal of industrial wastewaters.

Representative industrial wastes include: pulp and paper, slaughterhouse, brewery, tannery, food processing, cannery, chemical, petroleum, textile, sugar, laundry, meat and poultry, hog feeding, rendering and many others. The initial step in treatment design development is an industrial waste survey, which provides data on variations in flow and waste characteristics. Undesirable waste characteristics as listed by Eckenfelder can be summarized as follows:

- Soluble organics causing depletion of dissolved oxygen.
- Suspended solids.
- Trace organics.
- Heavy metals, cyanide and toxic organics.
- Colour and turbidity.
- Nitrogen and phosphorus.
- Refractory substances resistant to biodegradation.
- Oil and floating material.
- Volatile materials.

The US EPA has further defined a list of toxic organic and inorganic chemicals with specific limitations in granting discharge permits. The list includes more than 100 compounds and is too long to reprint here, but may be requested from the EPA.

Treatment Methods

The handling of industrial wastes is more specialized than the treatment of domestic wastes; however, where amenable to biological reduction, they are usually treated using methods similar to those previously described (secondary/tertiary biological treatment approaches) for municipal systems.

Waste stabilization ponds are a common method of organic wastewater treatment where sufficient land area is available. Flow-through ponds are generally classified according to their bacterial activity as aerobic, facultative or anaerobic. Aerated ponds are supplied with oxygen by diffused or mechanical aeration systems.

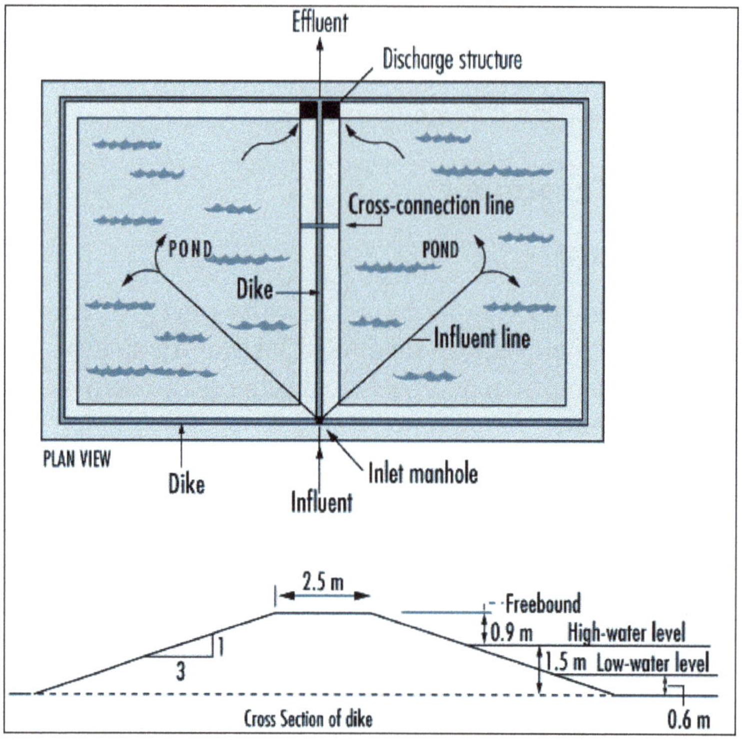

Two-cell stabilization pond: cross sectional diagram.

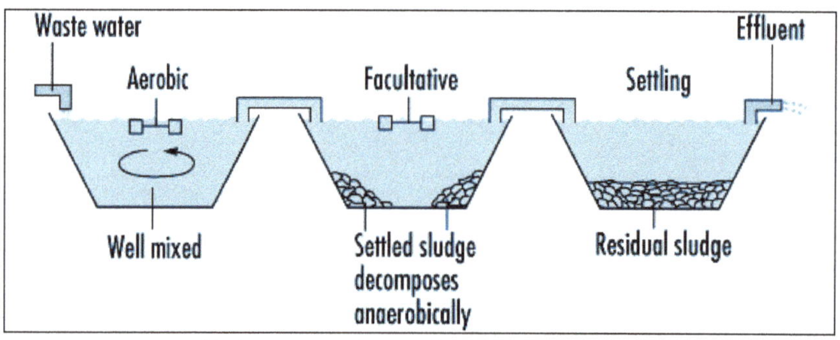

Pollution prevention and waste minimization.

When industrial waste in-plant operations and processes are analysed at their source, they often can be controlled so as to prevent significant polluting discharges. Recirculation techniques are important approaches in pollution prevention programmes.

Advanced Methods of Wastewater Treatment

A number of advanced methods are available for higher degrees of removal of pollution constituents as may be required. A general listing includes:

- Filtration (sand and multimedia),

- Chemical precipitation,
- Carbon adsorption,
- Electrodialysis,
- Distillation,
- Nitrification,
- Algae harvesting,
- Reclamation of effluents,
- Micro-straining,
- Ammonia stripping,
- Reverse osmosis,
- Ion exchange,
- Land application,
- Denitrification,
- Wetlands.

Thermal Pollution Control

The main approaches for the control of thermal pollution are:

- Improved power plant operation efficiencies.
- Cooling towers.
- Isolated cooling ponds.
- Consideration of alternative methods of power generation such as hydro-power.

Where physical conditions are favourable within certain environmental limits, hydro electric pow er should be considered as an alternative to fossil-fuel or nuclear power generation. In hydro-electric power generation, there is no disposal of heat and there is no discharge of waste waters causing water pollution.

Groundwater Pollution Control

Importance of Groundwater

Since the world's water supplies are widely extracted from aquifers, it is most important that these sources of supply be protected. It is estimated that more than 95% of the earth's available fresh water supply is underground; in the United States approximately 50% of the drinking water comes from wells, according to the 1984 US Geological Survey. Because underground water pollution and movement are of subtle and unseen nature, less attention sometimes is given to the analysis and control of this form of water degradation than to surface water pollution, which is far more obvious.

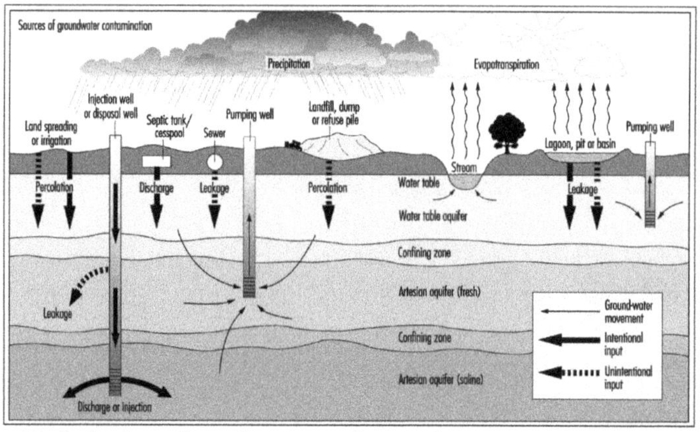

Hydrologic cycle and sources of groundwater contamination.

Underground Pollution

Figure above shows the hydrologic cycle with superimposed sources of groundwater contamination. A complete listing of the potential sources of underground pollution is extensive; however, for illustration the most obvious sources include:

- Industrial waste discharges.
- Polluted streams in contact with aquifers.
- Mining operations.
- Solid and hazardous waste disposal.
- Underground storage tanks such as for petroleum.
- Irrigation systems.
- Artificial recharge.
- Sea water encroachment.
- Spills.
- Polluted ponds with permeable bottoms.
- Disposal wells.
- Septic tank tile fields and leaching pits.
- Improper well drilling.
- Agricultural operations.
- Roadway de-icing salts.

Specific pollutants in underground contamination are further categorized as:

- Undesirable chemical constituents (typical, not complete list) - Organic and inorganic (e.g., chloride, sulphate, iron, manganese, sodium, potassium).
- total hardness and total dissolved solids.
- Toxic constituents (typical, not complete list) - Nitrate, arsenic, chromium, lead, cyanide, copper, phenols, dissolved mercury.
- Undesirable physical characteristics - Taste, colour and odour.
- Pesticides and herbicides - Chlorinated hydrocarbons and others.
- Radioactive materials - Various forms of radioactivity.
- Biological - Bacteria, viruses, parasites and soon.
- Acid (low pH) or caustic (high pH).

Of the above, nitrates are of special concern in both ground waters and surface waters. In groundwater supplies, nitrates can cause the disease methaemoglobinaemia (infant cyanosis). They further cause detrimental eutrophication effects in surface waters and occur in a wide range of water resources, as reported by Preul. Preul and Preul and Schroepfer have also reported on the underground movement of nitrogen and other pollutants.

Pollution Travel in Underground Domain

Groundwater movement is exceedingly slow and subtle as compared with the travel of surface waters in the hydrologic cycle. For a simple understanding of the travel of ordinary groundwater under ideal steady flow conditions, Darcy's Law is the basic approach for the evaluation of groundwater movement at low Reynolds numbers (R):

$$V = K(dh/dl)$$

where:

V = velocity of groundwater in aquifer, m/day

K = coefficient of permeability of aquifer

(dh/dl) = hydraulic gradient which represents the driving force for movement.

In pollutant travel underground, ordinary groundwater (H_2O) is generally the carrying fluid and can be calculated to move at a rate according to the parameters in Darcy's Law. However, the rate of travel or velocity of a pollutant, such as an organic or inorganic chemical, may be different due to advection and hydrodynamic dispersion processes. Certain ions move slower or faster than the general rate of groundwater flow as a result of reactions within the aquifer media, so that they can be categorized as "reacting" or "non-reacting". Reactions are generally of the following forms:

- Physical reactions between the pollutant and the aquifer and the transporting liquid.

- Chemical reactions between the pollutant and the aquifer and the transporting liquid.
- Biological actions on the pollutant.

The following are typical of reacting and non-reacting underground pollutants:

- Reacting pollutants - Chromium, ammonium ion, calcium, sodium, iron and so on; cations in general; biological constituents; radioactive constituents.
- Non-reacting pollutants - Chloride, nitrate, sulphate and so on; certain anions; certain pesticide and herbicide chemicals.

At first, it might seem that reacting pollutants are the worst type, but this may not always be the case because the reactions detain or retard pollutant travel concentrations whereas non-reacting pollutant travel may be largely uninhibited. Certain "soft" domestic and agricultural products are now available which biologically degrade after a period of time and therefore avoid the possibility of groundwater contamination.

Aquifer Remediation

Prevention of underground pollution is obviously the best approach; however, uncontrolled existence of polluted groundwater conditions usually is made known after its occurrence, such as by complaints from water well users in the area. Unfortunately, by the time the problem is recognized, severe damage may have occurred and remediation is necessary. Remediation may require extensive hydro-geological field investigations with laboratory analyses of water samples in order to establish the extent of pollutant concentrations and travel plumes. Often existing wells can be used in initial sampling, but severe cases may require extensive borings and water samplings. These data can then be analysed to establish current conditions and to make future condition predictions. The analysis of groundwater contamination travel is a specialized field often requiring the use of computer models to better understand the groundwater dynamics and to make predictions under various constraints. A number of two- and three-dimensional computer models are available in the literature for this purpose. For more detailed analytical approaches, the reader is referred to the book by Freeze and Cherry.

Pollution Prevention

The preferred approach for the protection of groundwater resources is pollution prevention. Although drinking water standards generally apply to the use of groundwater supplies, the raw water supplies require protection from contamination. Governmental entities such as ministries of health, natural resources agencies, and environmental protection agencies are generally responsible for such activities. Groundwater pollution control efforts are largely directed at protection of aquifers and the prevention of pollution.

Pollution prevention requires land-use controls in the form of zoning and certain regulations. Laws may apply to the prevention of specific functions as particularly applicable to point sources or actions which potentially may cause pollution. Control by land-use zoning is a groundwater protection tool which is most effective at the municipal or county level of government. Aquifer and wellhead protection programmes as discussed below are leading examples of pollution prevention.

An aquifer protection programme requires establishing the boundaries of the aquifer and its recharge areas. Aquifers may be of an unconfined or confined type, and therefore need to be analysed by a hydrologist to make this determination. Most major aquifers are generally well known in developed countries, but other areas may require field investigations and hydrogeologic analysis. The key element of the programme in the protection of the aquifer from water quality degradation is control of land use over the aquifer and its recharge areas.

Wellhead protection is a more definitive and limited approach which applies to the recharge area contributing to a particular well. The US federal government by amendments passed in 1986 to the Safe Drinking Water Act (SDWA) now requires that specific wellhead protection areas be established for public supply wells. The wellhead protection area (WHPA) is defined in the SDWA as "the surface and subsurface area surrounding a water well or well field, supplying a public water supply system, through which contaminants are reasonably likely to move toward and reach such water well or well field." The main objective in the WHPA programme, as outlined by the US EPA, is the delineation of well protection areas based on selected criteria, well operation and hydrogeologic considerations.

References

- Water-pollution, science: britannica.com, Retrieved 16 January, 2019
- Gary Stricker (28 July 1998). "Pollution invades small Pacific island". CNN. Archived from the original on 31 December 2007. Retrieved 1 April 2008
- 9-things-that-causes-water-pollution-and-its-possible-solution, water-filter: theberkey.com, Retrieved 29 March, 2019
- "Facts about marine debris". US NOAA. Archived from the original on 13 February 2009. Retrieved 10 April 2008
- Duce, R A; et al. (2008). "Impacts of Atmospheric Anthropogenic Nitrogen on the Open Ocean". Science. 320 (5878): 893–89. Bibcode:2008Sci...320..893D. Doi:10.1126/science.1150369. PMID 18487184
- Edgar B. Herwick III (29 July 2015). "Explosive Beach Objects-- Just Another Example of Massachusetts' Charm". WGBH news. PBS. Archived from the original on 3 August 2015. Retrieved 4 August2015
- Marine-pollution-causes-types-effects-prevention, society, my-india: mapsofindia.com, Retrieved 2 June, 2019
- Schindler, David and Vallentyne, John R. (2004) Over fertilization of the World's Freshwaters and Estuaries, University of Alberta Press, p. 1, ISBN 0-88864-484-1
- 511-water-pollution-control, item, environmental-pollution-control, part-vii: iloencyclopaedia.org, Retrieved 23 February, 2019

4
Land Pollution

The deposition of solid or liquid waste materials on land or underground is known as land pollution. It may contaminate the soil and groundwater along with threatening public health. The topics elaborated in this chapter will help in gaining a better understanding of land pollution as well as its causes.

Land pollution is the deposition of solid or liquid waste materials on land or underground in a manner that can contaminate the soil and groundwater, threaten public health, and cause unsightly conditions and nuisances.

The waste materials that cause land pollution are broadly classified as municipal solid waste (MSW, also called municipal refuse), construction and demolition (C&D) waste or debris, and hazardous waste. MSW includes nonhazardous garbage, rubbish, and trash from homes, institutions (e.g., schools), commercial establishments, and industrial facilities. Garbage contains moist and decomposable (biodegradable) food wastes (e.g., meat and vegetable scraps); rubbish comprises mostly dry materials such as paper, glass, textiles, and plastic objects; and trash includes bulky waste materials and objects that are not collected routinely for disposal (e.g., discarded mattresses, appliances, pieces of furniture). C&D waste (or debris) includes wood and metal objects, wallboard, concrete rubble, asphalt, and other inert materials produced when structures are built, renovated, or demolished. Hazardous wastes include harmful and dangerous substances generated primarily as liquids but also as solids, sludges, or gases by various chemical manufacturing companies, petroleum refineries, paper mills, smelters, machine shops, dry cleaners, automobile repair shops, and many other industries or commercial facilities. In addition to improper disposal of MSW, C&D waste, and hazardous waste, contaminated effluent from subsurface sewage disposal (e.g., from septic tanks) can also be a cause of land pollution.

The permeability of soil formations underlying a waste-disposal site is of great importance with regard to land pollution. The greater the permeability, the greater the risks from land pollution. Soil consists of a mixture of unconsolidated mineral and rock fragments (gravel, sand, silt, and clay) formed from natural weathering processes. Gravel and sand formations are porous and permeable, allowing the free flow of water through the pores or spaces between the particles. Silt is much less permeable than sand or gravel, because of its small particle and pore sizes, while clay is virtually impermeable to the flow of water, because of its platelike shape and molecular forces.

Until the mid-20th century, solid wastes were generally collected and placed on top of the ground in uncontrolled "open dumps" which often became breeding grounds for rats, mosquitoes, flies, and other disease carriers and were sources of unpleasant odours, windblown debris, and other nuisances. Dumps can contaminate groundwater as well as pollute nearby streams and lakes. A highly contaminated liquid called leachate is generated from decomposition of garbage and precipitation that infiltrates and percolates downward through the volume of waste material. When leachate reaches and mixes with groundwater or seeps into nearby bodies of surface water, public health and environmental quality are jeopardized. Methane, a poisonous and explosive gas that easily flows through soil, is an eventual by-product of the anaerobic (in the absence of oxygen) decomposition of putrescible solid waste material. Open dumping of solid waste is no longer allowed in many countries. Nevertheless, leachate and methane from old dumps continue to cause land pollution problems in some areas.

A modern technique for land disposal of solid waste involves construction and daily operation and control of so-called sanitary landfills. Sanitary landfills are not dumps; they are carefully planned and engineered facilities designed to control leachate and methane and minimize the risk of land pollution from solid-waste disposal. Sanitary landfill sites are carefully selected and prepared with impermeable bottom liners to collect leachate and prevent contamination of groundwater. Bottom liners typically consist of flexible plastic membranes and a layer of compacted clay. The waste material—MSW and C&D debris—is spread out, compacted with heavy machinery, and covered each day with a layer of compacted soil. Leachate is collected in a network of perforated pipes at the bottom of the landfill and pumped to an on-site treatment plant or nearby public sewerage system. Methane is also collected in the landfill and safely vented to the atmosphere or recovered for use as a fuel known as biogas, or landfill gas. Groundwater-monitoring wells must be placed around the landfill and sampled periodically to ensure proper landfill operation. Completed landfills are capped with a layer of clay or an impermeable membrane to prevent water from entering. A layer of topsoil and various forms of vegetation are placed as a final cover. Completed landfills are often used as public parks or playgrounds.

Hazardous waste differs from MSW and C&D debris in both form and behaviour. Its disposal requires special attention because it can cause serious illnesses or injuries and can pose immediate and significant threats to environmental quality. The main characteristics of hazardous waste include toxicity, reactivity, ignitability, and corrosivity. In addition, waste products that may be infectious or are radioactive are also classified as hazardous waste. Although land disposal of hazardous waste is not always the best option, solid or containerized hazardous wastes can be disposed of by burial in "secure landfills," while liquid hazardous waste can be disposed of underground in deep-well injection systems if the geologic conditions are suitable. Some hazardous wastes such as dioxins, PCBs, cyanides, halogenated organics, and strong acids are banned from land disposal in the United States, unless they are first treated or stabilized or meet certain concentration limits. Secure landfills must have at least 3 metres (10 feet) of soil between the bottom of the landfill and underlying bedrock or groundwater table (twice that required for municipal solid-waste landfills), a final impermeable cover when completed, and a double impervious bottom liner for increased safety. Underground injection wells (into which liquid waste is pumped under high pressure) must deposit the liquid in a permeable layer of rock that is sandwiched between impervious layers of rock or clay. The wells must also be encased and sealed in three concentric pipes and be at least 400 metres (0.25 mile) from any drinking-water supplies for added safety.

Before modern techniques for disposing of hazardous wastes were legislated and put into practice, the wastes were generally disposed of or stored in surface piles, lagoons, ponds, or unlined landfills. Thousands of those waste sites still exist, now old and abandoned. Also, the illegal but frequent practice of "midnight dumping" of hazardous wastes, as well as accidental spills, has contaminated thousands of industrial land parcels and continues to pose serious threats to public health and environmental quality. Efforts to remediate or clean up such sites will continue for years to come. In 1980 the United States Congress created the Superfund program and authorized billions of dollars toward site remediation; today there are still about 1,300 sites on the Superfund list requiring remediation. The first listed Superfund site—Love Canal, located in Niagara Falls, N.Y.—was not removed from the list until 2004.

Causes of Land Pollution

There are many different ways of permanently changing the land, from soil contamination (poisoning by chemicals or waste) to general urbanization (the systematic creation of cities and other human settlements from greenfield, virgin land). Some, such as huge landfills or quarries, are very obvious; others, such as atmospheric deposition (where land becomes contaminated when air pollution falls onto it) are much less apparent. Let's consider the main causes and types of land pollution in turn.

Waste Disposal

Humans produce vast quantities of waste—in factories and offices, in our homes and schools, and in such unlikely places as hospitals. Even the most sophisticated waste processing plants, which use plasma torches (electrically controlled "flames" at temperatures of thousands of degrees) to turn waste into gas, produce solid waste products that have to be disposed of somehow. There's simply no getting away from waste: our ultimate fate as humans is to die and become waste products that have to be burned or buried.

Chart showing the ordinary, hazardous, medical, and ash waste produced per person per year.

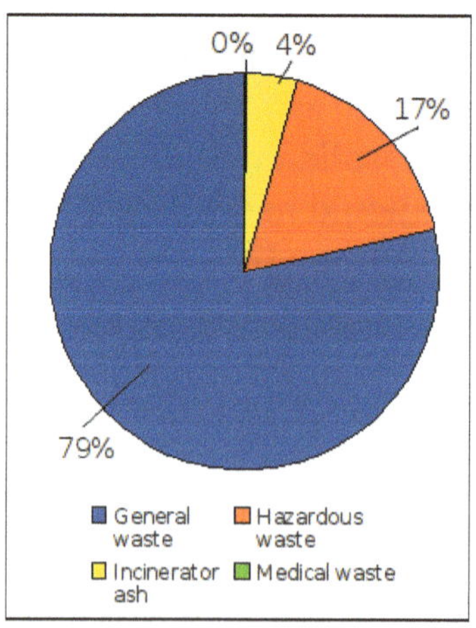

Chart: Although most of the waste we produce is relatively harmless and easy to dispose of (blue), around one fifth of it (orange, yellow, and green) is dangerous or toxic and extremely difficult to get rid of without automatically contaminating land.

Waste disposal didn't always mean land pollution. Before the 20th century, most of the materials people used were completely natural (produced from either plants, animals, or minerals found in the Earth) so, when they were disposed of, the waste products they generated were natural and harmless too: mostly organic (carbon-based) materials that would simply biodegrade (break down eventually into soil-like compost). There was really nothing we could put into the Earth that was more harmful than anything we'd taken from it in the first place. But during the 20th century, the development of plastics (polymers generally made in chemical plants from petroleum and other chemicals), composites (made by combining two or more other materials), and other synthetic (human-created) materials has produced a new generation of unnatural materials that the natural environment has no idea how to break down. It can take 500 years for a plastic bottle to biodegrade, for example. And while it's easy enough to recycle simple things such as cardboard boxes or steel cans, it's much harder to do the same thing with computer circuit boards made from dozens of different electronic components, themselves made from countless metals and other chemicals, all tightly bonded together and almost impossible to dismantle.

Nothing illustrates the problem of waste disposal more clearly than radioactive waste. When scientists discovered how to create energy by splitting atoms in nuclear power plants, they also created the world's hardest waste disposal problem. Nuclear plants produce toxic waste that can remain dangerously radioactive for thousands of years and, what's worse, will contaminate anything or anyone that comes into contact with it. Nuclear plants that have suffered catastrophic accidents are generally sealed with concrete and abandoned indefinitely. Not surprisingly, local communities object vociferously to having nuclear waste stored anywhere near them.

Mining

Satellite photo of Escondida Mine Chile.

Although there are many responsible mining companies, and environmental laws now tightly restrict mining in some countries, mines remain among the most obvious scars on (and under) the landscape. Surface mining (sometimes called quarrying or opencast mining) requires the removal of topsoil (the fertile layer of soil and organic matter that is particularly valuable for agriculture) to get at the valuable rocks below. Even if the destruction of topsoil is the worst that happens, it can turn a productive landscape into a barren one, which is a kind of pollution. You might think a mine would only remove things from the land, causing little or no pollution, but mining isn't so simple. Most metals, for example, occur in rocky mixtures called ores, from which the valuable elements have to be extracted by chemical, electrical, or other processes. That leaves behind waste products and the chemicals used to process them, which historically were simply dumped back on the land. Since all the waste was left in one place, the concentration of pollution often became dangerously high. When mines were completely worked out, all that was left behind was contaminated land that couldn't be used for any other purpose. Often old mines have been used as landfills, adding the insult of an inverted garbage mountain to the injury of the original damage. But at least it saved damaging more land elsewhere.

Urbanization

Humans have been making permanent settlements for at least 10,000 years and, short of some major accident or natural disaster, most of the cities and towns we've created, and the infrastructure that keeps them running, will remain with us for thousands more years into the future. Not many of us would automatically classify cities and other human settlements as "land pollution"; people obviously need to live and work somewhere. Even so, urbanization marks a hugely important change to the landscape that can cause land pollution in a variety of subtle and not-so-subtle ways.

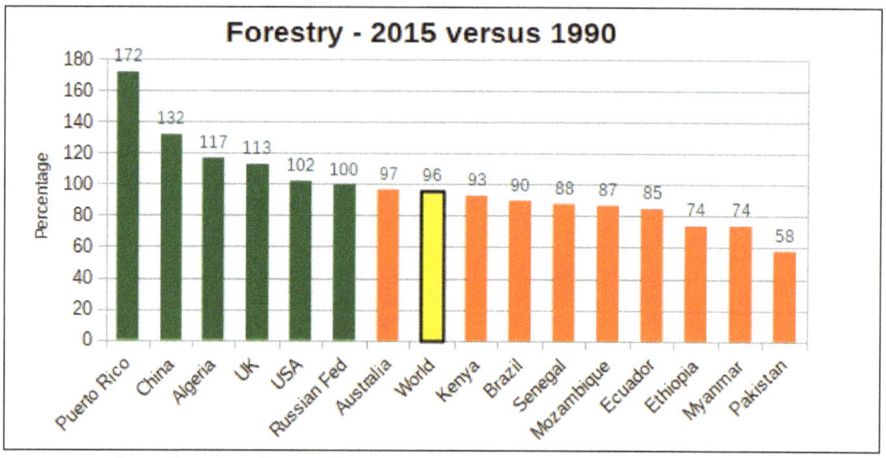

Bar chart showing forested area for 15 countries, 2015 compared to 1990.

Chart: Urbanization goes hand-in-hand with other changes in land use, such as deforestation. In 2015, the world had about 96 percent as much forested area as it had in 1990—a huge loss of forest in total. This chart shows 15 example countries that have either gained forest (green) or lost it (orange), with the world total shown in the middle (yellow). For each country, the bar shows the percentage of forest area in 2015 compared to 1990, so 100 percent would be no change. Drawn by explainthatstuff.com using data from UN Food and Agriculture Organization/World Bank, pulished under a Creative Commons BY-4.0 license.

With over 7.7 billion people on the planet, it might come as a surprise to find that humans have urbanized only about 3 percent of Earth's total land surface, though about 30–40 percent of the total land area has been transformed if we include agriculture. Our impact on the planet extends much further than urbanization might suggest. Way back in 1996, Herbert Girardet estimated that

London, England has an ecological footprint (area of land needed to support it) some 125 times bigger than the city itself. Add up that effect for every major city in the world and you get an idea of how big an impact urbanization has had. Today's figures are staggering. According to the Global Footprint Network, the ecological footprint of most countries (what they use) hugely exceeds their biocapacity (what they can produce): in the United States, the ecological footprint per person is 2.3 times bigger than the biocapacity; in Germany it's 2.8 times bigger; in China, 3.7 times bigger; and in India, 2.2 times bigger.

One of the problems of urbanization is that, by concentrating people, it concentrates their waste products at the same time. So, for example, crudely disposing of sewage from a big city automatically creates water or land pollution, where the same number of people and the same volume of sewage might not create a problem if it were created in 10 smaller cities or 100 small towns. Concentration is always a key factor when we talk about pollution. Having said that, it's important to remember that urbanization, when it works, can also help people to live very efficiently. Thus, New York has the lowest ecological footprint of any state in the USA, largely because people there have smaller homes and make greater use of public transportation.

Orange bulldozers turning a green field into mud for construction of houses.

Agricultural Chemicals

Those of us who are lucky enough to live in rich countries take our basic survival for granted: aside from trips to the grocery store, we don't worry about where our food comes from or how it gets to us. The reality is that seven billion hungry people consume a vast amount of food. Feeding the world on such a scale is only possible because agriculture now works in an industrial way, with giant machines such as tractors and combine harvesters doing the work that hundreds of people would have done in the past, and chemicals such as fertilizers and pesticides (herbicides that kill weeds and insecticides that kill bugs) increasing the amount of food that can be grown on each piece of land. Unfortunately, most pesticides are by definition poisons, and many remain in the soil or accumulate there for years. One infamous and now widely banned pesticide, DDT, is not ordinarily biodegradable so it has remained in the environment ever since it was first used in the mid-20th century and even spread to such places as Antarctica. DDT is just one of many organic (carbon-based) chemicals that remain in the environment for years or decades, known as persistent organic pollutants.

Atmospheric Deposition

Air pollution doesn't remain air pollution forever. Ideally it disperses, so the concentration of problematic chemicals becomes so low that it no longer constitutes pollution. Sometimes, though, it falls back to the ground and becomes either water pollution (if it enters the oceans, rivers, and lakes) or land pollution. Pollution created ("deposited") in water or land from existing pollution in the air (atmosphere) is known as atmospheric deposition. Land can become polluted by deposition in some very unexpected ways. For example, a corridor of land either side of a highway or freeway becomes systematically polluted over time with all kinds of harmful byproducts of road travel—everything from fuel spills and brake linings to dust worn from the pavement and heavy metal deposits (such as lead) washed from the engines. These chemicals accumulate in the soil where they can undergo reactions with one another and form substances that are even more toxic.

Two important things are worth noting about atmospheric deposition. First, it means no land on Earth—not even the most isolated island—can be considered completely safe from pollution: even if it's hundreds or thousand miles from the nearest factory or human settlement, even if no human has ever lived there, it could still be polluted from the air. Second, if you're doing something that causes pollution (maybe spreading weedkiller on your garden or perhaps running a factory where ash is discharged from a smokestack), the effects are not necessarily going to be confined to the place where the pollution is first produced. It's important to remember that pollution knows no boundaries.

Soil Erosion

An agricultural scientist examines soil that has turned to dust in Mexico due to soil erosion by the wind.

Soil erosion turns fields into deserts.

If you define "land pollution" as irreversible damage to the land, you have to include soil erosion as a type of pollution too. Many people think soil is soil, always there, never changing, ever ready to grow whatever crops we choose to bury in it. In reality, soil is a much more complex growing habitat that remains productive only when it is cared for and nurtured. Too much wind or water, destruction of soil structure by excessive plowing, excessive nutrients, overgrazing, and overproduction of crops erode soil, damaging its structure and drastically reducing its productivity until it's little more than dust. At its worst, soil erosion becomes desertification: once-productive agricultural areas become barren, useless deserts. How serious is the problem? In 2001, former UN Secretary General Kofi Annan warned the world that: "Drought and desertification threaten the livelihood of over 1 billion people in more than 110 countries around the world." Deforestation doesn't only harm the place where the trees are cut down. A 2013 study by Princeton University researchers found that if the Amazon rainforest were completely destroyed, it would have a dramatic effect on the atmosphere, which would carry across to places like the United States, causing drought and potentially desertification there as well.

Unfortunately, because soil erosion has so far affected developing countries more than the developed world, it's a problem that receives relatively little attention. Accelerating climate change will soon alter that. In a future of hotter weather and more intense storms, it will become increasingly difficult to maintain soil in a fertile and productive state, while heavy rainstorms and flash floods will wash away topsoil more readily. Meanwhile, agriculture may become impossible in coastal areas inundated by saltwater carried in by rising sea levels. We might think of global warming as an example of air pollution (because it's caused mostly by humans releasing gases such as carbon dioxide into the atmosphere). But if it leads to dramatic sea-level rise and coastal erosion, you could argue that it will become an example of land pollution as well.

Industry, Energy, Manufacturing and Construction Waste

Industrial wastes that pollute land include paints, chemicals, plastics, and metals among other industrial manufacturing byproducts and residues. The major industrial wastes products are generated from power plants, oil refineries, construction works, pharmaceuticals, and agricultural product producers.

Energy producing power plants release chemical wastes that are disposed in landfills. Examples include nuclear, coal, and fossil fuel driven energy production processes. Oil refinery processes also produce petroleum hydrocarbon byproducts that end up as waste and construction sites generate metal, wood, and plastic wastes whereas agricultural industries release scores of chemical wastes that find their way into landfills.

Power plants combust fossil fuels and biomass while other industrial manufacturers such as pharmaceuticals, oil refineries, and agricultural product producers utilize a series of raw materials with a lot of chemical residues and byproducts. As much as the bulk of industrial waste products disposals are regulated, at times, they find way to the landfills or end up disposed on some land somewhere. Wastewater and liquids from industrial, agricultural farms and manufacturing processes also contain all kinds of chemicals that at times contaminate lands when they are not disposed off correctly.

Oil Spill

We need refined oil to run our vehicles. Crude oil is refined into usable petrol, gas or diesel. When it is done, it results in some byproducts. Those byproducts end up at waste and are buried deep underground or in the landfills.

Similarly, leaking oil from vehicles can contaminate the soil and can prove serious for human health if same oil is used for agricultural activities.

Ashes

Ashes refer to the residual matter that is left behind after burning solid fuels. Generally, solid wastes like glass and metals from hospitals, market places, offices, and public places among other waste collection points are collected and burned in incinerators. After incineration, the remaining matter is ash, and they are categorized into two.

The first residual type is the bottom ash that remains at the bottom and second is called fly ash that is trapped by filters in the smokestack of the incinerators. Fly ash is highly toxic. About every four trucks of waste burnt produce one truck of ashes that is eventually dumped into the landfills.

Natural Disasters

Natural disasters in the form of floods, hurricanes, tornadoes, cyclones, earthquakes can destroy buildings and carry hazardous chemicals and materials into waterways and oceans. These pollutants can disrupt marine life and aquatic systems.

Radioactive Waste

This includes waste from nuclear power stations that is disposed off underground. Rays from this kind of waste can cause lung or skin cancer. Radon is a serious pollutant gas that appears naturally in soil as a result of uranium breakdown process. This can have hazardous effect on the health of humans and wildlife.

Deforestation

Deforestation is the cut down of trees to clear land for farming, construction, settlement, mining, or other economic purposes. When this happens, it results in both land degradation and pollution because the value of land is completely lost. The naturally existing vegetation cover is responsible for supporting ecosystems and various life forms.

In forest regions, for instance, trees absorb and reflect 20% of the sun's heat, conserving and protecting the surface soils from adverse land degradation or desertification. When trees are cut, the earth's surface is left unprotected from rain and sun's intense heat. Accordingly, the unfavorable conditions result in soil erosion, land degradation, and desertification.

Effects of Land Pollution

The pollution of land has calamitous consequences especially concerning the survival of animals and humans and the quality of soil and water. Environmental protection agencies indicate that the effects can even be more devastating if wastes in landfills are not separated into reusable, recyclable, or organic waste. Since the advent of industrialization, human activities have gradually destroyed and degraded land, causing diseases to humans and animals and reducing the capacity to support ecosystem and various life forms. Land pollution has many long lasting effects. Below are the main consequences of land pollution:

Heavy machinery at the Robin Hood Bay landfill.

Effect on Human Health

Many regions across the world harbor loads of waste and have incredibly pitiable waste collection services especially in developing countries and slum areas. Others have appropriate waste collection systems, but they eventually end up in the landfills without proper regulation.

Such wastes contain dangerous chemicals, pesticides, and metals that have adverse effects on humans. Plastic waste, for instance, might contain acrylic, polyvinyl chloride, polycarbonate, and phthalates that are associated with cancers, skin diseases, respiratory disorders, and birth defects for pregnant women.

Chemical components such as cadmium, asbestos, mercury, cyanide, arsenic, and chromium commonly found in pharmaceutical, pesticides, and fertilizer industrial wastes also have devastating effects on human health. They have cancer causing elements and can also lead to lung and kidney disease as well as liver damage.

Increase in Landfill Sites

When land is contaminated with solid agricultural and industrial waste, it leads to increase in landfill sites across the city. Landfills also become breeding grounds for mice, rodents, flies, and birds that can transmit diseases. These landfills are contaminated with such kind of toxic chemicals that they can reach the human body via vegetables and foods that are grown in polluted lands. They can also seep into water bodies used for consumer purposes or could be inhaled by humans from polluted dust. Apart form that, these huge pile of wastes across the city results in tourist distraction and can be loss of revenue for tourism industry.

Soil Pollution

Soil pollution is categorized under land pollution. Hence, when there is overuse of fertilizer chemicals or lands are degraded through chemical and solid waste dumping, the upper layer of the soil is damaged, causing soil pollution. The effects of agricultural, healthcare, and chemical wastes are the principal land pollution aspects causing soil pollution.

Healthcare waste such as broken instruments and metals or industrial consumer product waste like broken electronics are all contaminated waste that may end up in landfills thereby destroying the soils and land on which they eventually wind up, upsetting the land ecosystems. Ultimately, the land losses its fertility and vegetation cover.

Air Pollution

Landfills and dump sites generate appalling smells and odour in the areas which they are located. In cities and towns located near huge dump sites and landfill areas, residents have experienced high scores of pungent smell. Apart from the bad smell, landfills are always burning which contributes to air pollution.

Water Pollution

Land pollution can spread in all directions so that it results in an adverse impact on the immediate environments. On this basis, it can contaminate water and significantly reduce its quality. It happens when the chemicals and other toxic substances from the landfills and solid wastes are mostly carried into waterways by surface rainwater runoff.

At the same time, leaching takes place which makes the toxic elements and chemicals to infiltrate into aquifers and water tables. Also, the contaminated water evaporates and falls back as precipitation with the impurities, advancing the cycle of pollution and contamination.

Environmental Degradation

Deforestation is the top concern for matters pertaining environmental degradation. It causes the destruction of the ecosystems and the habitat. Deforestation compromises the cutting down of vegetation and tree cover which creates harsh environmental conditions such as intense heat from the sun and also disrupts the rain cycle.

It results in soil erosion and desertification. As an outcome, animals are driven further away and become vulnerable to harmful conditions and predators. Continued deforestation finally results in biodiversity implications because the reduction of green cover interferes with the atmospheric balance.

When the atmospheric balance is compromised, various atmospheric imbalances such as the greenhouse effect, climate change, and global warming are experienced. Deforestation is as well a recipe for increased stormwater runoff and leaching. The economic effect of deforestation is depriving communities of revenue from tourism.

Effect on Wildlife

In the recent decades, the wildlife animals have tremendously suffered as they have persistently faced a serious threat regarding the loss of their natural habitat and environment. The continued human economic activities on land have progressively left the lands degraded and polluted forcing wildlife to move further away and adapt to new areas. Consequently, some species have died while trying to adapt, some have gone extinct, and several are currently on the verge of extinction.

Discharge of various kinds of chemicals on land makes the ecosystem unappealing to the survival of plants and animals within their interactive food chain. The chemicals contaminate plants and waters which are then consumed by lower animals, and the food chain continues up the ladder in the ecosystem. The process is known as biomagnification and is reported to be a serious threat to ecological sustainability.

Effect on Tourism Industry

Landfills and scattered dump sites within cities typically give a very bad picture of the residents and the governance of the respective cities. It may only depict the level of environmental insensitivity. Landfills and dump sites also reduce air quality and can potentially threaten human health. As such, it can make a city loss its attraction to tourists, in turn, making the government to loss tourism revenue.

Effect on Plant Life

When soils are repeatedly contaminated and accumulate large amounts of poisonous materials and chemicals, the soil reaches a point where it cannot support plant life. Soil pollutants interfere with soil chemistry, biology, and structure. When these changes occur, beneficial soil bacteria, soil microorganisms, soil nutrients, and soil chemical processes begin to deteriorate to an extent where they diminish soil fertility.

The ecological balance is lost completely. On this basis, the soil becomes unsuitable for crop survival or any other form of vegetation. If the plants die, then it means animals dependent on the plants will also die. This leads to migration of the larger animals and predators to other regions to find food supply, gradually leading to a reduction in wildlife and extinction.

Soil pollution can as well change plant metabolism and lower crop productivity. Besides, when plants take up the soil contaminants, they pass them up the food chain, endangering the health of animals and humans.

Acidification

Soil pollution allows emission of relatively large quantities of nitrogen via denitrification, volatilization of ammonia, and the decomposition of organic materials in the soil. As a result, this releases sulfur compounds and sulfur dioxides into the atmosphere, causing acid rain.

In the long-run, it leads to a continuous cycle of pollution whereby the acid rain reduces soil chemistry and nutrients, which would further contribute to ecological balance disturbance and soil erosion. Furthermore, acidic conditions hinder soil ability to cushion changes in the soil PH, leading to the death of plants due to unfavorable conditions.

Diminished Soil Fertility

The most evident and crucial element of the soil is its fertility. Once the soil is contaminated with chemicals and heavy metals or degraded due to human activities such as mining, its fertility depreciates and might even be lost entirely. The harmful chemicals and heavy metals in the soil decrease soil microbial and chemical activity.

The chemical elements can also denature active enzymes that revitalize healthy soil activities. Soil acidification as a consequence of pollution also leaches away essential natural minerals like magnesium and calcium.

Changes in the Soil Structure

Acidification, diminished soil fertility, and death of soil organisms in the soil can lead to changes in soil structure. This is because soil microorganisms help in breaking down organic matter that promotes soil structure regarding vitality and water penetration as well as retention.

Increase in Soil Salinity

The increase in soil salinity, salinization, is an effect of salt accumulation in the soil. Salts occur naturally in the soil. However, increased accumulations are linked to soil pollution. Irrigation and agricultural processes that discharge nitrate and phosphate deposits in the soil are the primary contributors to increasing salt levels in the soil.

Increased soil salinity makes it difficult for plants to absorb soil moisture and reduces groundwater quality. Crops and plants grown in these regions combined with other soil pollutant effects are highly poisonous and can cause severe health disorders when consumed.

References

- Land-pollution, science: britannica.com, Retrieved 10 July, 2019
- Land-pollution: explainthatstuff.com, Retrieved 13 March, 2019
- Causes-of-land-pollution, pollution: eartheclipse.com, Retrieved 29 June, 2019
- Serious-effects-of-land-pollution, pollution: eartheclipse.com, Retrieved 30 January, 2019
- Devastating-effects-of-soil-pollution, pollution: eartheclipse.com, Retrieved 31 March, 2019

5

Noise Pollution

The propagation of noise with harmful effects on the activity of human and animal lives is known as noise pollution. A few of the types of noise pollution are occupational noise and aircraft noise pollution. This chapter has been carefully written to provide an easy understanding of these types of noise pollution as well as its control and regulation.

Noise pollution is an unwanted or excessive sound that can have deleterious effects on human health and environmental quality. Noise pollution is commonly generated inside many industrial facilities and some other workplaces, but it also comes from highway, railway, and airplane traffic and from outdoor construction activities.

Measuring and Perceiving Loudness

Sound waves are vibrations of air molecules carried from a noise source to the ear. Sound is typically described in terms of the loudness (amplitude) and the pitch (frequency) of the wave. Loudness (also called sound pressure level, or SPL) is measured in logarithmic units called decibels (dB). The normal human ear can detect sounds that range between 0 dB (hearing threshold) and about 140 dB, with sounds between 120dB and 140 dB causing pain (pain threshold). The ambient SPL in a library is about 35 dB, while that inside a moving bus or subway train is roughly 85 dB; building construction activities can generate SPLs as high as 105 dB at the source. SPLs decrease with distance from the source.

The rate at which sound energy is transmitted, called sound intensity, is proportional to the square of the SPL. Because of the logarithmic nature of the decibel scale, an increase of 10 dB represents a 10-fold increase in sound intensity, an increase of 20 dB represents a 100-fold increase in intensity, a 30-dB increase represents a 1,000-fold increase in intensity, and so on. When sound intensity is doubled, on the other hand, the SPL increases by only 3 dB. For example, if a construction drill causes a noise level of about 90 dB, then two identical drills operating side by side will cause a noise level of 93 dB. On the other hand, when two sounds that differ by more than 15 dB in SPL are combined, the weaker sound is masked (or drowned out) by the louder sound. For example, if an 80-dB drill is operating next to a 95-dB dozer at a construction site, the combined SPL of those two sources will be measured as 95 dB; the less intense sound from the compressor will not be noticeable.

Frequency of a sound wave is expressed in cycles per second (cps), but hertz (Hz) is more commonly used (1 cps = 1 Hz). The human eardrum is a very sensitive organ with a large dynamic range, being able to detect sounds at frequencies as low as 20 Hz (a very low pitch) up to about 20,000 Hz (a very high pitch). The pitch of a human voice in normal conversation occurs at frequencies between 250 Hz and 2,000 Hz.

Precise measurement and scientific description of sound levels differ from most subjective human perceptions and opinions about sound. Subjective human responses to noise depend on both pitch and loudness. People with normal hearing generally perceive high-frequency sounds to be louder than low-frequency sounds of the same amplitude. For this reason, electronic sound-level meters used to measure noise levels take into account the variations of perceived loudness with pitch. Frequency filters in the meters serve to match meter readings with the sensitivity of the human ear and the relative loudness of various sounds. The so-called A-weighted filter, for example, is commonly used for measuring ambient community noise. SPL measurements made with this filter are expressed as A-weighted decibels, or dBA. Most people perceive and describe a 6- to 10-dBA increase in an SPL reading to be a doubling of "loudness." Another system, the C-weighted (dBC) scale, is sometimes used for impact noise levels, such as gunfire, and tends to be more accurate than dBA for the perceived loudness of sounds with low frequency components.

Noise levels generally vary with time, so noise measurement data are reported as time-averaged values to express overall noise levels. There are several ways to do this. For example, the results of a set of repeated sound-level measurements may be reported as L_{90} = 75 dBA, meaning that the levels were equal to or higher than 75 dBA for 90 percent of the time. Another unit, called equivalent sound levels (L_{eq}), can be used to express an average SPL over any period of interest, such as an eight-hour workday. (L_{eq} is a logarithmic average rather than an arithmetic average, so loud events prevail in the overall result.) A unit called day-night sound level (DNL or L_{dn}) accounts for the fact that people are more sensitive to noise during the night, so a 10-dBA penalty is added to SPL values that are measured between 10 PM and 7 AM. DNL measurements are very useful for describing overall community exposure to aircraft noise, for example.

Dealing with the effects of Noise

Noise is more than a mere nuisance. At certain levels and durations of exposure, it can cause physical damage to the eardrum and the sensitive hair cells of the inner ear and result in temporary or permanent hearing loss. Hearing loss does not usually occur at SPLs below 80 dBA (eight-hour exposure levels are best kept below 85 dBA), but most people repeatedly exposed to more than 105 dBA will have permanent hearing loss to some extent. In addition to causing hearing loss, excessive noise exposure can also raise blood pressure and pulse rates, cause irritability, anxiety, and mental fatigue, and interfere with sleep, recreation, and personal communication. Noise pollution control is therefore of importance in the workplace and in the community. Noise-control ordinances and laws enacted at the local, regional, and national levels can be effective in mitigating the adverse effects of noise pollution.

Environmental and industrial noise is regulated in the United States under the Occupational Safety and Health Act of 1970 and the Noise Control Act of 1972. Under these acts, the Occupational Safety and Health Administration set up industrial noise criteria in order to provide limits on the intensity of sound exposure and on the time duration for which that intensity may be allowed.

If an individual is exposed to various levels of noise for different time intervals during the day, the total exposure or dose (D) of noise is obtained from the relation:

$$D = (C_1/T_1) + (C_2/T_2) + (C_3/T_3) + \ldots,$$

where C is the actual time of exposure and T is the allowable time of exposure at any level. Using this formula, the maximum allowable daily noise dose will be 1, and any daily exposure over 1 is unacceptable.

Criteria for indoor noise are summarized in three sets of specifications that have been derived by collecting subjective judgments from a large sampling of people in a variety of specific situations. These have developed into the noise criteria (NC) and preferred noise criteria (PNC) curves, which provide limits on the level of noise introduced into the environment. The NC curves, developed in 1957, aim to provide a comfortable working or living environment by specifying the maximum allowable level of noise in octave bands over the entire audio spectrum. The complete set of 11 curves specifies noise criteria for a broad range of situations. The PNC curves, developed in 1971, add limits on low-frequency rumble and high-frequency hiss; hence, they are preferred over the older NC standard. Summarized in the curves, these criteria provide design goals for noise levels for a variety of different purposes. Part of the specification of a work or living environment is the appropriate PNC curve; in the event that the sound level exceeds PNC limits, sound-absorptive materials can be introduced into the environment as necessary to meet the appropriate standards.

Low levels of noise may be overcome using additional absorbing material, such as heavy drapery or sound-absorbent tiles in enclosed rooms. Where low levels of identifiable noise may be distracting or where privacy of conversations in adjacent offices and reception areas may be important, the undesirable sounds may be masked. A small white-noise source such as static or rushing air, placed in the room, can mask the sounds of conversation from adjacent rooms without being offensive or dangerous to the ears of people working nearby. This type of device is often used in offices of doctors and other professionals. Another technique for reducing personal noise levels is through the use of hearing protectors, which are held over the ears in the same manner as an earmuff. By using commercially available earmuff-type hearing protectors, a decrease in sound level can be attained ranging typically from about 10 dB at 100 Hz to more than 30 dB for frequencies above 1,000 Hz.

Outdoor noise limits are also important for human comfort. Standard house construction will provide some shielding from external sounds if the house meets minimum standards of construction and if the outside noise level falls within acceptable limits. These limits are generally specified for particular periods of the day—for example, during daylight hours, during evening hours, and at night during sleeping hours. Because of refraction in the atmosphere owing to the nighttime temperature inversion, relatively loud sounds can be introduced into an area from a rather distant highway, airport, or railroad. One interesting technique for control of highway noise is the erection of noise barriers alongside the highway, separating the highway from adjacent residential areas. The effectiveness of such barriers is limited by the diffraction of sound, which is greater at the lower frequencies that often predominate in road noise, especially from large vehicles. In order to be effective, they must be as close as possible to either the source or the observer of the noise (preferably to the source), thus maximizing the diffraction that would be necessary for the sound to reach the observer. Another requirement for this type of barrier is that it must also limit the amount of transmitted sound in order to bring about significant noise reduction.

Types of Noise Pollution

Occupational Noise

Occupational noise is the amount of acoustic energy received by an employee's auditory system when they are working in the industry. Occupational noise, or industrial noise, is often a term used in occupational safety and health, as sustained exposure can cause permanent hearing damage.

"Twenty-two million workers are exposed to potentially damaging noise at work each year. Last year, U.S. business paid more than $1.5 million in penalties for not protecting workers from noise." - OSHA.

Occupational noise is considered an occupational hazard traditionally linked to loud industries such as ship-building, mining, railroad work, welding, and construction, but can be present in any workplace where hazardous noise is present.

Regulation

In the United States, the National Institute for Occupational Safety and Health (NIOSH) and the Occupational Safety and Health Administration (OSHA) work together to provide standards and regulations for noise in the workplace.

National Institute for Occupational Safety and Health (NIOSH), Occupational Safety and Health Administration (OSHA), Mine Safety and Health Administration (MSHA), Federal Railroad Administration (FRA) have all set standards on hazardous occupational noise in their respective industries. Each industry is different, as workers' tasks and equipment differ, but most regulations agree that noise becomes hazardous when it exceeds 85 decibels, for an 8-hour time exposure (typical work shift). This relationship between allotted noise level and exposure time is known as an Exposure action value (EAV) or Permissible exposure limit (PEL). The EAV or PEL can be seen as equations which manipulate the allotted exposure time according to the intensity of the industrial noise. This equation works as an inverse, exponential, relationship. As the industrial noise intensity increases, the allotted exposure tie, to still remain safe, decreases.

The above calculations of PEL and EAV are based on measurements taken to determine the intensity of that particular industrial noise. A-weighted measurements are commonly used to determine noise levels that can cause harm to the human ear. There are also special exposure meters available that integrate noise over a period of time to give an Leq value (equivalent sound pressure level), defined by standards.

Risks of Occupational Hearing Loss

Occupational noise, if experienced repeatedly, at high intensity, for an extended period of time, can cause noise-induced hearing loss (NIHL) which is then classified as occupational hearing loss.

Noise, in the context of industrial noise, is hazardous to a person's hearing because of its loud intensity through repeated long-term exposure. In order for noise to cause hearing impairment for the worker, the noise has to be close enough, loud enough, and sustained long enough to damage the hair cells in the auditory system. These factors have been taken into account by the governing occupational health and safety organization to determine the unsafe noise exposure levels and durations for their respective industries.

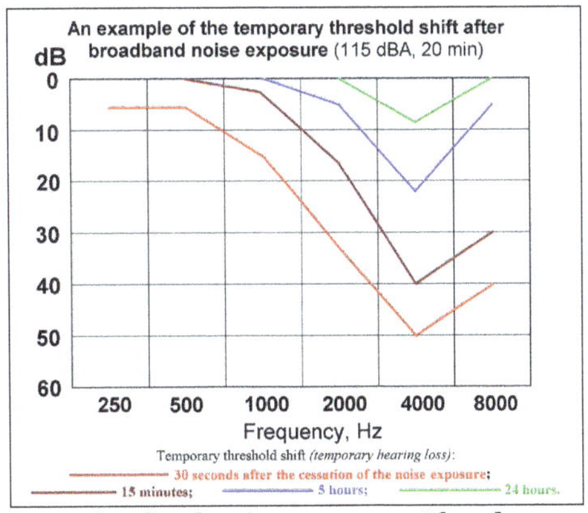

Temporary hearing loss after loud noise exposure. If such exposure is long enough, this temporary threshold shift may become permanent.

Noise can also affect the safety of the employee and others. Noise can be a causal factor in work accidents as it may mask hazards and warning signals and impede concentration. High intensity noise interferes with vital workplace communication which increases the chance of accidents and decreases productivity.

Noise may also act synergistically with other hazards to increase the risk of harm to workers. In particular, toxic materials (e.g. some solvents, metals, asphyxiants and pesticides) have some oto-toxic properties that may affect hearing function.

Modern thinking in occupational safety and health further identifies noise as hazardous to workers' safety and health. This hazard is experienced in various places of employment and through a variety of sources.

Noise, in the context of industrial noise, is hazardous to a persons hearing because of its loud intensity through repeated long-term exposure. In order for Noise to cause Hearing impairment for the worker, the noise has to be close enough, loud enough and the listener has to be exposed for long enough. These factors have been taken into account by the governing occupational health and safety organizations as they determine the unsafe noise exposure levels and durations for their respective industries.

National Institute for Occupational Safety and Health (NIOSH), Occupational Safety and Health Administration (OSHA), Mine Safety and Health Administration (MSHA), Federal Railroad Administration (FRA) have all set standards on hazardous occupational noise in their respective industries. Each industry is different, as workers tasks and equipment differ, but most regulations agree that noise becomes hazardous when it exceeds 85 Decibel, for an 8-hour exposure (typical work shift). This relationship between allotted noise level and exposure time is known as an Exposure action value (EAV) or Permissible exposure limit (PEL). The EAV or PEL can be seen as equations which manipulate the allotted exposure time according to the intensity of the industrial noise. This equation works as an inverse relationship. As the industrial noise intensity increases, the allotted exposure time, to still remain safe, decreases.

These above calculations of PEL and EAV are based on measurements taken to determine the intensity of that particular industrial noise. A-weighted measurements are commonly used to determine noise levels that can cause harm to the human ear. There are also special exposure meters available that integrate noise over a period of time to give an L_{eq} value (equivalent sound pressure level), defined by standards.

Reduction

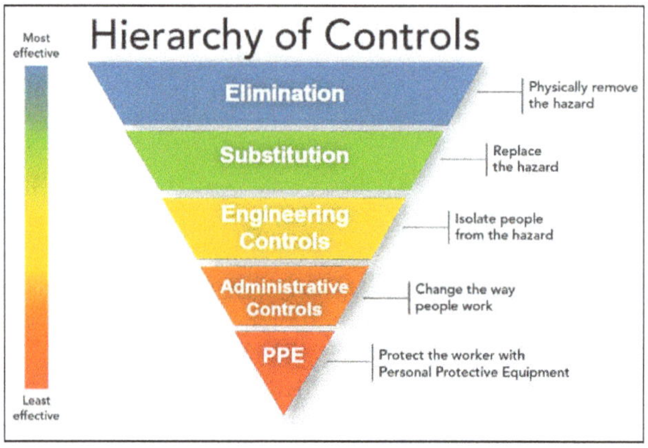

Hierarchy of Controls guideline.

There are several ways to limit your exposure to hazardous occupational noise. The hierarchy of controls is a guideline for reducing hazardous noise. First, the company can eliminate the noise source. If the noise source cannot be eliminated, the company must try to reduce the noise with alternative methods. This process is called acoustic quieting.

Acoustic quieting is the process of making machinery quieter by damping vibrations to prevent them from reaching the observer. The company can isolate the certain piece of machinery by placing materials on the machine or in between the machine and the worker to decreases the signal intensity that reaches the worker's ear.

To decrease an employee's exposure to hazardous noise, the company can also take administrative control by limiting the employee's exposure time. This can be done by changing work shifts and switching employees out from the noise exposure area. Lastly, to decrease occupational noise exposure, hearing protection should be used. There are several types of earplugs and earmuffs that can be used to attenuate the noise to a safe level.

Aircraft Noise Pollution

Aircraft noise pollution is a harmful noise effect produced by any aircraft or its components during the various phases of a flight. Sound production is divided into three categories:

- Mechanical noise—Rotation of the engine parts, most noticeable when fan blades reach supersonic speeds.
- Aerodynamic noise—From the airflow around the surfaces of the aircraft, especially when flying low at high speeds.
- Noise from aircraft systems—Cockpit and cabin pressurization and conditioning systems, and Auxiliary Power units.

Health consequences include sleep disturbance, hearing impairment and heart disease, as well as workplace accidents caused by stress. Memory and recall can also be affected. Governments have enacted extensive controls that apply to aircraft designers, manufacturers, and operators, resulting in improved procedures and cuts in pollution.

Mechanisms of Sound Production

Small general aviation aircraft produce localized aircraft noise.

Helicopter main and tail rotors produce aerodynamic noise.

Aircraft noise is noise pollution produced by an aircraft or its components, whether on the ground while parked such as auxiliary power units, while taxiing, on run-up from propeller and jet exhaust, during take off, underneath and lateral to departure and arrival paths, over-flying while en route, or during landing. A moving aircraft including the jet engine or propeller causes compression and rarefaction of the air, producing motion of air molecules. This movement propagates through the air as pressure waves. If these pressure waves are strong enough and within the audible frequency spectrum, a sensation of hearing is produced. Different aircraft types have different noise levels and frequencies. The noise originates from three main sources:

- Engine and other mechanical noise.
- Aerodynamic noise.
- Noise from aircraft systems.

Engine and other Mechanical Noise

Much of the noise in propeller aircraft comes equally from the propellers and aerodynamics. Helicopter noise is aerodynamically induced noise from the main and tail rotors and mechanically induced noise from the main gearbox and various transmission chains. The mechanical sources produce narrow band high intensity peaks relating to the rotational speed and movement of the moving parts. In computer modelling terms noise from a moving aircraft can be treated as a line source.

Aircraft gas turbine engines (jet engines) are responsible for much of the aircraft noise during takeoff and climb, such as the buzzsaw noise generated when the tips of the fan blades reach supersonic speeds. However, with advances in noise reduction technologies—the airframe is typically more noisy during landing.

The majority of engine noise is due to jet noise—although high bypass-ratio turbofans do have considerable fan noise. The high velocity jet leaving the back of the engine has an inherent shear layer instability (if not thick enough) and rolls up into ring vortices. This later breaks down into turbulence. The SPL associated with engine noise is proportional to the jet speed (to a high power). Therefore, even modest reductions in exhaust velocity will produce a large reduction in jet noise.

Engines are the main source of aircraft noise. The geared Pratt & Whitney PW1000G helped reduce the noise levels of the Bombardier CSeries, Mitsubishi MRJ and Embraer E-Jet E2 crossover narrowbody aircraft: the gearbox allows the fan to spin at an optimal speed, which is one third the speed of the LP turbine, for slower fan tip speeds. It has a 75% smaller noise footprint than current equivalents. The PowerJet SaM146 in the Sukhoi Superjet 100 features 3D aerodynamic fan blades and a nacelle with a long mixed duct flow nozzle to reduce noise.

Aerodynamic Noise

Aerodynamic noise arises from the airflow around the aircraft fuselage and control surfaces. This type of noise increases with aircraft speed and also at low altitudes due to the density of the air. Jet-powered aircraft create intense noise from aerodynamics. Low-flying, high-speed military aircraft produce especially loud aerodynamic noise.

The shape of the nose, windshield or canopy of an aircraft affects the sound produced. Much of the noise of a propeller aircraft is of aerodynamic origin due to the flow of air around the blades. The helicopter main and tail rotors also give rise to aerodynamic noise. This type of aerodynamic noise is mostly low frequency determined by the rotor speed.

Typically noise is generated when flow passes an object on the aircraft, for example, the wings or landing gear. There are broadly two main types of airframe noise:

- Bluff Body Noise – the alternating vortex shedding from either side of a bluff body, creates low-pressure regions (at the core of the shed vortices) which manifest themselves as pressure waves (or sound). The separated flow around the bluff body is quite unstable, and the flow "rolls up" into ring vortices—which later break down into turbulence.
- Edge Noise – when turbulent flow passes the end of an object or gaps in a structure (high lift device clearance gaps) the associated fluctuations in pressure are heard as the sound propagates from the edge of the object (radially downwards).

Noise from Aircraft Systems

Cockpit and cabin pressurization and conditioning systems are often a major contributor within cabins of both civilian and military aircraft. However, one of the most significant sources of cabin noise from commercial jet aircraft, other than the engines, is the Auxiliary Power Unit (APU), an onboard generator used in aircraft to start the main engines, usually with compressed air, and to provide electrical power while the aircraft is on the ground. Other internal aircraft systems can also contribute, such as specialized electronic equipment in some military aircraft.

Health Effects

Aircraft engines are the major source of noise and can exceed 140 decibels (dB) during takeoff. While airborne, the main sources of noise are the engines and the high speed turbulence over the fuselage.

There are health consequences of elevated sound levels. Elevated workplace or other noise can cause hearing impairment, hypertension, ischemic heart disease, annoyance, sleep disturbance, and decreased school performance. Although some hearing loss occurs naturally with age, in many developed nations the impact of noise is sufficient to impair hearing over the course of a lifetime. Elevated noise levels can create stress, increase workplace accident rates, and stimulate aggression and other anti-social behaviors. Airport noise has been linked to high blood pressure.

German Environmental Study

A large-scale statistical analysis of the health effects of aircraft noise was undertaken in the late 2000s by Bernhard Greiser for the Umweltbundesamt, Germany's central environmental office. The health data of over one million residents around the Cologne airport were analysed for health effects correlating with aircraft noise. The results were then corrected for other noise influences in the residential areas, and for socioeconomic factors, to reduce possible skewing of the data.

The German study concluded that aircraft noise clearly and significantly impairs health. For example, a day-time average sound pressure level of 60 decibels increasing coronary heart disease by 61% in men and 80% in women. As another indicator, a night-time average sound pressure level of 55 decibels increased the risk of heart attacks by 66% in men and 139% in women. Statistically significant health effects did however start as early as from an average sound pressure level of 40 decibels.

FAA Advice

The Federal Aviation Administration (FAA) regulates the maximum noise level that individual civil aircraft can emit through requiring aircraft to meet certain noise certification standards. These standards designate changes in maximum noise level requirements by "stage" designation. The U.S. noise standards are defined in the Code of Federal Regulations (CFR) Title 14 Part 36 – Noise Standards: Aircraft Type and Airworthiness Certification (14 CFR Part 36). The FAA says that a maximum day-night average sound level of 65 dB is incompatible with residential communities. Communities in affected areas may be eligible for mitigation such as soundproofing.

Cabin Noise

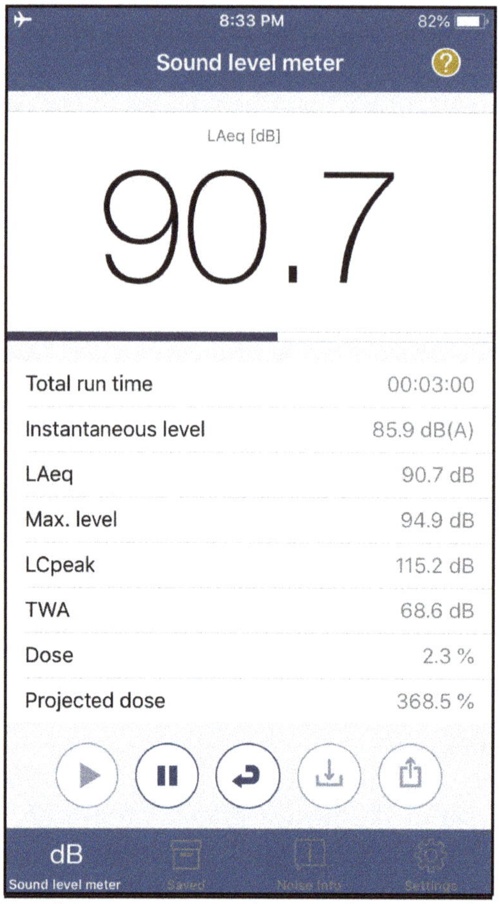

NIOSH Sound Level Meter app averaging 90.7 dB at takeoff in a Boeing 777.

Aircraft noise also affects people within the aircraft: crew and passengers. Cabin noise can be studied to address the occupational exposure and the health and safety of pilots and flight attendants. In 1998, 64 commercial airline pilots were surveyed regarding hearing loss and tinnitus. In 1999, the NIOSH conducted several noise surveys and health hazard evaluations, and found noise levels exceeding its recommended exposure limit of 85 A-weighted decibels as an 8-hr TWA. In 2006, the noise levels inside an Airbus A321 during cruise have been reported as approximately 78 dB(A) and during taxi when the aircraft engines are producing minimal thrust, noise levels in the cabin have been recorded at 65 dB(A). In 2008, a study of Swedish airlines cabin crews found average sound levels between 78–84 dB(A) with maximum A-weighted exposure of 114 dB but found no major hearing threshold shifts. In 2018, a study of sound levels measured on 200 flights representing six aircraft groups found media noise level of 83.5 db(A) with levels reaching 110 dB(A) on certain flights, but only 4.5% exceeded the NIOSH recommended 8-hr TWA of 85 dB(A).

Cognitive Effects

Simulated aircraft noise at 65 dB(A) has been shown to negatively affect individuals' memory and recall of auditory information. In one study comparing the effect of aircraft noise to the effect of alcohol on cognitive performance, it was found that simulated aircraft noise at 65 dB(A) had the same effect on individuals' ability to recall auditory information as being intoxicated with a Blood

Alcohol Concentration (BAC) level of at 0.10. A BAC of 0.10 is double the legal limit required to operate a motor vehicle in many developed countries such as Australia.

Health Effects from Noise

Noise health effects are the physical and psychological health consequences of regular exposure to consistent elevated sound levels. Elevated workplace or environmental noise can cause hearing impairment, tinnitus, hypertension, ischemic heart disease, annoyance, and sleep disturbance. Changes in the immune system and birth defects have been also attributed to noise exposure.

Although age-related health effects (presbycusis) occur naturally with age, in many countries the cumulative impact of noise is sufficient to impair the hearing of a large fraction of the population over the course of a lifetime. Noise exposure has been known to induce tinnitus, hypertension, vasoconstriction, and other cardiovascular adverse effects. Chronic noise exposure has been associated with sleep disturbances and increased incidence of diabetes. Adverse cardiovascular effects occur from chronic exposure to noise due to the sympathetic nervous system's inability to habituate. The sympathetic nervous system maintains lighter stages of sleep when the body is exposed to noise, which does not allow blood pressure to follow the normal rise and fall cycle of an undisturbed circadian rhythm.

Stress from time spent around elevated noise levels has been linked with increased workplace accident rates and aggression and other anti-social behaviors. The most significant sources are vehicles, aircraft, prolonged exposure to loud music, and industrial noise. There are an 10,000 deaths per year as a result of noise in the European Union.

Noise Induced Hearing Loss

Noise-induced hearing loss is a permanent shift in pure-tone thresholds, resulting in sensorineural hearing loss. The severity of a threshold shift is dependent on duration and severity of noise exposure. Noise-induced threshold shifts are seen as a notch on an audiogram from 3000–6000 Hz, but most often at 4000 Hz.

Exposure to loud noises, either in a single traumatic experience or over time, can damage the auditory system and result in hearing loss and sometimes tinnitus as well. Traumatic noise exposure can happen at work (e.g. loud machinery), at play (e.g. loud sporting events, concerts, recreational activities), and/or by accident (e.g. a backfiring engine.) Noise induced hearing loss is sometimes unilateral and typically causes patients to lose hearing around the frequency of the triggering sound trauma.

Tinnitus

Tinnitus is an auditory disorder characterized by the perception of a sound (ringing, chirping, buzzing, etc.) in the ear in the absence of an external sound source. There are two types of tinnitus: subjective and objective. Subjective is the most common and can only be heard "in the head" by the person affected. Objective tinnitus can be heard from those around the affected person.

Though the pathophysiology of tinnitus isn't known, noise exposure can be a contributing factor. Noise-induced tinnitus can be temporary or permanent depending on the type and amount of noise a person was exposed to.

Cardiovascular Effects

Noise has been associated with important cardiovascular health problems, particularly hypertension. Noise levels of 50 dB(A) at night may also increase the risk of myocardial infarction by chronically elevating cortisol production. Roadway noise levels are sufficient to constrict arterial blood flow and lead to elevated blood pressure. Vasoconstriction can result from elevated adrenaline levels or through medical stress reactions.

Psychological Impacts of Noise

Causal relationships have been discovered between noise and psychological effects such as annoyance, psychiatric disorders, and effects on psychosocial well-being. Exposure to intense levels of noise can cause personality changes and violent reactions. Noise has also been shown to be a factor that attributed to violent reactions. The psychological impacts of noise also include an addiction to loud music. This was researched in a study where non-professional musicians were found to have loudness addictions more often than non-musician control subjects.

Psychological health effects from noise include depression and anxiety. Individuals who have hearing loss, including noise induced hearing loss, may have their symptoms alleviated with the use of hearing aids. Individuals who do not seek treatment for their loss are 50% more likely to have depression than their aided peers. These psychological effects can lead to detriments in physical care in the form of reduced self-care, work-tolerance, and increased isolation.

Auditory stimuli can serve as psychological triggers for individuals with post traumatic stress disorder (PTSD).

Stress

Research commissioned by Rockwool, a multi-national insulation manufacturer headquartered in Denmark, reveals that in the UK one third (33%) of victims of domestic disturbances claim loud parties have left them unable to sleep or made them stressed in the last two years. Around one in eleven (9%) of those affected by domestic disturbances claims it has left them continually disturbed and stressed. More than 1.8 million people claim noisy neighbours have made their life a misery and they cannot enjoy their own homes. The impact of noise on health is potentially a significant problem across the UK given that more than 17.5 million Britons (38%) have been disturbed by the inhabitants of neighbouring properties in the last two year. For almost one in ten (7%) Britons this is a regular occurrence.

The extent of the problem of noise pollution for public health is reinforced by figures collated by Rockwool from local authority responses to a Freedom of Information Act (FOI) request. This research reveals in the period April 2008 - 2009 UK councils received 315,838 complaints about noise pollution from private residences. This resulted in environmental health officers across the UK serving 8,069 noise abatement notices, or citations under the terms of the Anti-Social Behaviour (Scotland) Act.

Westminster City Council has received more complaints per head of population than any other district in the UK with 9,814 grievances about noise, which equates to 42.32 complaints per thousand residents. Eight of the top 10 councils ranked by complaints per 1,000 residents are located in London.

Annoyance

Sudden impulse noises are typically perceived as more bothersome than noise from traffic of equal volume. Annoyance effects of noise are minimally affected by demographics, but fear of the noise source and sensitivity to noise both strongly affect the 'annoyance' of a noise. Sound levels as low as 40 dB(A) can generate noise complaints and the lower threshold for noise producing sleep disturbance is 45 dB(A) or lower.

Other factors that affect the 'annoyance level' of sound include beliefs about noise prevention and the importance of the noise source, and annoyance at the cause (i.e. non-noise related factors) of the noise. Many of the interpretations of the level of annoyance and the relationship between noise levels and resulting health symptoms could be influenced by the quality of interpersonal relationships at the workplace, as well as the stress level generated by the work itself. Evidence for impact on annoyance of long-term noise versus recent changes is equivocal.

Approximately 35% to 40% of office workers find noise levels from 55 to 60 dB(A) extremely irritating. The noise standard in Germany for mentally stressful tasks is set at 55 dB(A), however, if the noise source is continuous, the threshold level for tolerability among office workers is lower than 55 dB(A).

Child Physical Development

The U.S. Environmental Protection Agency authored a pamphlet in 1978 that suggested a correlation between low-birthweight (using the World Health Organization definition of less than 2,500 grams (88 oz) and high sound levels, and also high rates of birth defects in places where expectant mothers are exposed to elevated sound levels, such as typical airport environs. Specific birth abnormalities included harelip, cleft palate, and defects in the spine.

According to Lester W. Sontag of the Fels Research Institute (as presented in the same EPA study): "There is ample evidence that environment has a role in shaping the physique, behavior, and function of animals, including man, from conception and not merely from birth. The fetus is capable of perceiving sounds and responding to them by motor activity and cardiac rate change." The effects of noise exposure are highest when it occurs between 15 and 60 days after conception, a period in which major internal organs and the central nervous system are formed.

Later developmental effects occur as vasoconstriction in the mother reduces blood flow and therefore oxygen and nutrition to the fetus. Low birth weights and noise were also associated with lower levels of certain hormones in the mother. These hormones are thought to affect fetal growth and to be good indicators of protein production. The difference between the hormone levels of pregnant mothers in noisy versus quiet areas increased as birth approached.

In a 2000 publication, a review of studies on birthweight and noise exposure note that while some older studies suggest that when women are exposed to >65 dB aircraft noise a small decrease in birthweight occurs, in a more recent study of 200 Taiwanese women including noise dosimetry measurements of individual noise exposure, the authors found no significant association between noise exposure and birth weight after adjusting for relevant confounders, e.g. social class, maternal weight gain during pregnancy, etc.

Cognitive Development

When young children are regularly exposed to levels of noise that interfere with speech, they may develop speech or reading difficulties, because auditory processing functions are compromised. Children continue to develop their speech perception abilities until they reach their teens. Evidence has shown that when children learn in noisier classrooms, they have more difficulties understanding speech than those who learn in quieter settings.

In a study conducted by Cornell University in 1993, children exposed to noise in learning environments experienced trouble with word discrimination, as well as various cognitive developmental delays. In particular, the writing learning impairment dysgraphia is commonly associated with environmental stressors in the classroom.

High noise levels have also been known to damage the physical health of small children. Children from noisy residences often have a heart rate that is significantly higher (by 2 beats/min on average) than those of children from quieter homes.

Prevention

Different styles of earplugs are pictured. Left, pre-molded earplugs. Center, formable earplugs. Right, roll-down foam earplugs.

A hearing protection device (HPD) is an ear protection device worn in or over the ears while exposed to hazardous noise to help prevent noise-induced hearing loss. HPDs reduce (not eliminate) the level of the noise entering the ear. HPDs can also protect against other effects of noise exposure such as tinnitus and hyperacusis. Proper hygiene and care of HPDs may reduce chances of outer ear infections. There are many different types of HPDs available for use, including earmuffs, earplugs, electronic hearing protection devices, and semi-insert devices. One can measure the personal attenuation rating through a hearing protection fit-testing system.

Earmuff style hearing protection devices are designed to fit over the outer ear, or pinna. Earmuff HPDs typically consist of two ear cups and a head band. Earplug style hearing protection devices are designed to fit in the ear canal. Earplugs come in a variety of different subtypes. Some HPDs reduce the sound reaching the eardrum through a combination of electronic and structural components. Electronic HPDs are available in both earmuff and custom earplug styles. Electronic microphones, circuitry, and receivers perform active noise reduction, also known as noise-cancelling, in which a signal that is 180-degrees out-of-phase of the noise is presented, which in theory cancels the noise. Canal caps are similar to earplugs in that they consists of soft tip that is inserted into the opening of the ear canal.

Regulations

Environmental noise regulations usually specify a maximum outdoor noise level of 60 to 65 dB(A), while occupational safety organizations recommend that the maximum exposure to noise is 40 hours per week at 85 to 90 dB(A). For every additional 3 dB(A), the maximum exposure time is reduced by a factor 2, e.g. 20 hours per week at 88 dB(A). Sometimes, a factor of two per additional 5 dB(A) is used, however, these occupational regulations are acknowledged by the health literature as inadequate to protect against hearing loss and other health effects. In an effort to prevent noise-induced hearing loss, many programs and initiative have been created, like the Buy Quiet program, which encourages employers to purchase quieter tools and equipment, and the Safe-In-Sound Award, which recognizes organizations with successful hearing loss prevention strategies.

With regard to indoor noise pollution in residences, the U.S. Environmental Protection Agency (EPA) has not set any restrictions on limits to the level of noise. Rather, it has provided a list of recommended levels in its Model Community Noise Control Ordinance, which was published in 1975. For instance, the recommended noise level for indoor residences is less than or equal to 45 dB.

Noise pollution control in residences is not funded by the federal government in part because of the disagreements in establishing causal links between sounds and health risks, since the effect of noise is often psychological and also, because it leaves no singular tangible trace of damage on the human body. For instance, hearing loss could be attributed to a variety of factors including age, rather than solely due to excessive exposure to noise. A state or local government is able to regulate indoor residential noise, however, such as when excessive noise from within a home causes disturbances to nearby residences.

Effects on Canines

While people are often educated on the effects of noise exposure in humans, there are also different noise exposure effects in animals as well. An example of this would be in canines, and the noise exposure levels occurring within kennels. Canines experience this noise exposure whether it be a long stay at an animal shelter, or a weekend stay at a boarding facility.

Organizations like NIOSH and OSHA have different regulations when it comes to the noise exposure levels in industrial workers. Currently there are no regulations related to the noise exposure in canines even with such damaging effects related to their health. Health risks dogs are exposed to include ear damage and behavioral changes.

The average noise exposure in a kennel is greater than 100 dB SPL. According to OSHA these levels would yield in the use of hearing protection for the workers of those kennels due to the risk of noise induced hearing loss. The anatomical structures of the human and canine ear are very similar, so it is thought that these levels will negatively impact the hearing of canines in kennels. The ABR can be used to estimate the hearing threshold of canines, and can be used to show either a temporary threshold shift or permanent threshold shift after being exposed to excessive sound levels.

Behavioral effects to excessive noise exposure include hiding, urinating, defecating, panting, pacing, drooling, disregard to commands, trembling, and barking. These behavioral patterns pose a much greater problem to canines than meets the eye. All of these behavioral patterns are characteristics that result in a longer stay at the kennels before being adopted. A longer stay at the shelter results in a longer duration of noise exposure and therefore more likely to show either a temporary or permanent threshold shift in the canine's hearing.

These excessive noise levels are not only harming the canines' physical and psychological state, but the workers' and potential adoptive families' physical and psychological state as well. The workers' psychological state could affect the care provided to the canines. These loud noise exposures also have the potential to reduce the amount of time that potential adoptive families spend in the facility. This can result in less dogs being adopted and more time being exposed to excessive sound levels.

To reduce the level of noise exposure poses a little more difficulty because the majority of the noise is coming from the canines (barking), but structural changes can be made to the facilities in order to reduce the noise. Structural changes could include how many dogs are put in one area, more absorbing material rather than metal cages and cement walls and floors, and possibly in the future use of hearing protection devices (HPD) for the canines. All of these structural changes would also benefit the humans involved as well as the use of HPD's (ear plugs).

Noise-induced Hearing Loss

Noise-induced hearing loss (NIHL) is hearing impairment resulting from exposure to loud sound. People may have a loss of perception of a narrow range of frequencies or impaired perception of sound including sensitivity to sound or ringing in the ears. When exposure to hazards such as noise occur at work and is associated with hearing loss, it is referred to as occupational hearing loss.

Hearing may deteriorate gradually from chronic and repeated noise exposure (such as to loud music or background noise) or suddenly from exposure to impulse noise, which is a short high intensity noise (such as a gunshot or airhorn). In both types, loud sound overstimulates delicate hearing cells, leading to the permanent injury or death of the cells. Once lost this way, hearing cannot be restored in humans.

There are a variety of prevention strategies available to avoid or reduce hearing loss. Lowering the volume of sound at its source, limiting the time of exposure and physical protection can reduce the impact of excessive noise. If not prevented, hearing loss can be managed through assistive devices and communication strategies.

The largest burden of NIHL has been through occupational exposures; however, noise-induced hearing loss can also be due to unsafe recreational, residential, social and military service-related noise exposures. It is estimated that 15% of young people are exposed to sufficient leisure noises (i.e. concerts, sporting events, daily activities, personal listening devices, etc.) to cause NIHL. There is not a limited list of noise sources that can cause hearing loss; rather, exposure to excessively high levels from any sound source over time can cause hearing loss.

Signs and Symptoms

The first symptom of NIHL may be difficulty hearing a conversation against a noisy background. The effect of hearing loss on speech perception has two components. The first component is the loss of audibility, which may be perceived as an overall decrease in volume. Modern hearing aids compensate this loss with amplification. The second component is known as "distortion" or "clarity loss" due to selective frequency loss. Consonants, due to their higher frequency, are typically affected first. For example, the sounds "s" and "t" are often difficult to hear for those with hearing loss, affecting clarity of speech. NIHL can affect either one or both ears. Unilateral hearing loss causes problems with directional hearing, affecting the ability to localize sound.

Temporary and Permanent Hearing Changes

- PTS (Permanent Threshold Shift) is a permanent change of the hearing threshold (the intensity necessary for one to detect a sound) following an event, which will never recover. PTS is measured in decibels.
- TTS (Temporary Threshold Shift) is a temporary change of the hearing threshold the hearing loss that will be recovered after a few hours to couple of days. Also called auditory fatigue. TTS is also measured in decibels.

In addition to hearing loss, other external symptoms of an acoustic trauma can be:

- Tinnitus,
- Otalgia,
- Hyperacusis,
- Dizziness or vertigo; in the case of vestibular damages, in the inner-ear.

Tinnitus

Tinnitus is described as hearing sound when an external sound is not present. Noise-induced hearing loss can cause high-pitched tinnitus. An estimated 50 million Americans have some degree of tinnitus in one or both ears; 16 million of them have symptoms serious enough for them to see a doctor or hearing specialist. As many as 2 million become so debilitated by the unrelenting ringing, hissing, chirping, clicking, whooshing or screeching, that they cannot carry out normal daily activities.

Tinnitus is the largest single category for disability claims in the military, with hearing loss a close second. The third largest category is post-traumatic stress disorder, which itself may be accompanied by tinnitus and may exacerbate it.

Quality of Life

NIHL has implications on quality of life that extend beyond related symptoms and the ability to hear. The annual disability-adjusted life years (DALYs) were estimated for noise-exposed U.S. workers. DALYs represent the number of healthy years lost due to a disease or other health condition. They were defined by the 2013 Global Burden of Disease (GBD) Study. The DALYs calculation accounts for life limitations experienced because of hearing loss as a lost portion of a healthy year of life. So the results indicate the number of healthy years lost by a group of people over a specific time period.

NIOSH used DALYs to estimate the impact of hearing loss on quality of life in the CDC Morbidity and Mortality Weekly Report article "Hearing Impairment Among Noise-Exposed Workers in the United States, 2003-2012." It reported that 2.5 healthy years were lost each year for every 1,000 noise-exposed U.S. workers because of hearing impairment (hearing loss that impacts day-to-day activities). These lost years were shared among the 13% of workers with hearing impairment (about 130 workers out of each 1,000 workers). Mining, Construction and Manufacturing workers lost more healthy years than workers in other industry sectors; specifically and respectively in those sectors, 3.5, 3.1 and 2.7 healthy years were lost each year for every 1,000 workers.

Negative Impacts

The negative impacts of NIHL on one's ability to reciprocate communication, socialize and interact with society are largely invisible. Hearing loss, in general, is not just an issue of volume; individuals may experience difficulty in understanding what is said over the phone, when several people are talking at once, in a large space, or when the speaker's face cannot be seen. Subsequently, challenging social interactions can negatively lead to decreased self-esteem, shame, and fear. This can be more acutely felt by those who experience hearing impairment or loss earlier in life, rather than later when it is more socially accepted. Such psychosocial states, regardless of age, can lead to social isolation, which is known to negatively impact one's overall health and well-being. The compounding impacts can also lead to depression, especially if hearing impairment leads to tinnitus. Research suggests that those with hearing impairment or loss may be at a greater risk for deterioration of quality of life, as captured by a quote from Helen Keller: "Blindness cuts us off from things, but deafness cuts us off from people." Hearing impairment and loss of hearing, regardless of source or age, also limits experiencing the many benefits of sound on quality of life. In addition to the interpersonal social benefits, new studies suggest the effects of nature sounds, such as birds chirping and water, can positively affect an individual's capacity to recover after being stressed or to increase cognitive focus.

Quality of Life Questionnaire

Hearing loss is typically quantified by results from an audiogram; however, the degree of loss of hearing does not predict the impact on one's quality of life. The impact that NIHL can have on daily life and psychosocial function can be assessed and quantified using a validated questionnaire tool, such as the Hearing Handicap Inventory for the Elderly (HHIE). The HHIE is considered a "useful tool for quantifying the perceived emotional and social/situational consequences of hearing loss." The original tool was designed to test adults 65 years of age and older; however, modified versions exist. For adults the Hearing Handicap Inventory for Adults (HHIA) can be used and for adolescents the modified 28-item Hearing Environments And Reflection on Quality of Life (HEAR-QL-28) can be used. The HHIA, for example, is a 25-item questionnaire that asks both social and emotional-specific questions such as: Does a hearing problem cause you to avoid groups of people?" (social) and "Does a hearing problem cause you to feel frustrated when talking to members of your family?" (emotional). Response options are yes, no and sometimes. A greater score indicates greater perceived handicap.

Cause

The ear can be exposed to short periods of sound in excess of 120 dB without permanent harm — albeit with discomfort and possibly pain; but long term exposure to sound levels over 85 dB(A) can cause permanent hearing loss.

There are two basic types of NIHL:

- NIHL caused by acoustic trauma,
- Gradually developing NIHL.

Acute Acoustic Trauma

NIHL caused by acute acoustic trauma refers to permanent cochlear damage from a one-time exposure to excessive sound pressure. This form of NIHL commonly results from exposure to high-intensity sounds such as explosions, gunfire, a large drum hit loudly, and firecrackers. According to one US study, excessive noise levels in cinemas are sufficiently brief that movie-goers do not experience hearing loss.

Perceived Harmfulness vs. Actual Harmfulness

The discomfort threshold is the loudness level from which a sound starts to be felt as too loud and thus painful by an individual. Industry workers tend to have a higher discomfort threshold (i.e. the sounds must be louder to feel painful than for non-industry workers), but the sound is just as harmful to their ears. Industry workers often suffer from NIHL because the discomfort threshold is not a relevant indicator of the harmfulness of a sound.

Gradually Developing

Gradually developing NIHL refers to permanent cochlear damage from repeated exposure to loud sounds over a period of time. Unlike acoustic trauma, this form of NIHL does not occur from a single exposure to a high-intensity sound pressure level. Gradually developing NIHL can be caused by multiple exposures to excessive noise in the workplace or any source of repetitive, frequent exposures to sounds of excessive volume, such as home and vehicle stereos, concerts, nightclubs, and personal media players.

Workplace

About 22 million workers are exposed to hazardous noise, with additional millions exposed to solvents and metals that could put them at increased risk for hearing loss. Occupational hearing loss is one of the most common occupational diseases. 49% of male miners have hearing loss by the age of 50. By the age of 60, this number goes up to 70%. Construction workers also suffer an elevated risk. A screening program focused on construction workers employed at US Department of Energy facilities found 58% with significant abnormal hearing loss due to noise exposures at work. Occupational hearing loss is present in up to 33% of workers overall. Occupational exposure to noise causes 16% of adult disabling hearing loss worldwide.

The following is a list of occupations that are most susceptible to hearing loss:

- Agriculture

- Mining
- Construction
- Manufacturing
- Utilities
- Transportation
- Military
- Musicians
- Orchestra conductors

Among Musicians

Musicians, from classical orchestras to rock groups, are exposed to high decibel ranges. Some rock musicians experience noise-induced hearing loss from their music, and some studies have found that "symphonic musicians suffer from hearing impairment and that the impairment might be ascribed to symphonic music."

Music-induced hearing loss is still a controversial topic for hearing researchers. While some population studies have shown that the risk for hearing loss increases as music exposure increases, other studies found little to no correlation between the two. Experts at the 2006 "Noise-Induced Hearing Loss in Children at Work and Play" Conference agreed that further research into this field was still required before making a broad generalization about music-induced hearing loss.

Given the extensive research suggesting that industrial noise exposure can cause sensorineural hearing loss, a link between hearing loss and music exposures of similar level and duration (to industrial noise) seems highly plausible. Determining which individuals or groups are at risk for such exposures may be a difficult task. Despite concerns about the proliferation of personal music players, the there is only scarce evidence supporting their impact on hearing loss, and some small-sample studies suggest that only a fraction of users are affected. People from ages to 6–19 have an approximately 15% rate of hearing loss. Recommendations for musicians to protect their hearing were released in 2015 by NIOSH. The recommendations emphasized education of musicians and those who work in or around the music industry. Annual hearing assessments were also recommended to monitor thresholds as were sound level assessments to help determine the amount of time musicians and related professionals should spend in that environment. Hearing protection was also recommended, and the authors of the NIOSH recommendations even suggested that musicians consider custom earplugs as a way to combat NIHL. Despite these recommendations, musicians continue to face unique challenges in protecting their hearing when compared to individuals in industrial settings. Typically environmental controls are the first line of defense in a hearing conservation program, and depending on the type of musicians some recommendations have been set in place.

These recommendations can include adjusting risers or level of the speakers and adjusting the layout of a band or orchestra. These changes in the environment can be beneficial to musicians, but the ability to accomplish them is not always possible. In the cases where these changes cannot be made, hearing protection is recommended. Hearing protection in musicians offers its own sets of benefits and complications. When used properly, hearing protection can limit the exposure of noise in individuals. Musicians have the ability to pick from several different types of hearing protection from conventional ear plugs to custom or high fidelity hearing protection. Despite this, use of hearing protection among musicians is low for several different reasons. Musicians often feel that hearing protective devices can distort how music sounds or makes it too quiet for them to hear important cues which makes them less likely to wear hearing protection even when aware of the risks. Research suggests that educations programs can be beneficial to musicians as well as working with hearing care professionals to help address the specific issues that musicians face.

In 2018, a musician named Chris Goldscheider won a case against Royal Opera House for damaging his hearing in a rehearsal of Wagner's thunderous opera Die Walkure.

Workplace Standards

In the United States, the Occupational Safety and Health Administration (OSHA) describes standards for occupational noise exposure in articles 1910.95 and 1926.52. OSHA states that an employer must implement hearing conservation programs for employees if the noise level of the workplace is equal to or above 85 dB(A) for an averaged eight-hour time period. OSHA also states that "exposure to impulsive or impact noise should not exceed 140 dB peak sound pressure level". The National Institute for Occupational Safety and Health (NIOSH) recommends that all worker exposures to noise should be controlled below a level equivalent to 85 dBA for eight hours to minimize occupational noise induced hearing loss. NIOSH also recommends a 3 dBA exchange rate so that every increase by 3 dBA doubles the amount of the noise and halves the recommended amount of exposure time. The United States Department of Defense (DoD) instruction 605512 has some differences from OSHA 1910.95 standard, for example, OSHA 1910.95 uses a 5 dB exchange rate and DoD instruction 605512 uses a 3 dB exchange rate.

There are programs that seek to increase compliance and therefore effectiveness of hearing protection rules; the programs include the use of hearing tests and educating people that loud sound is dangerous. Employees are required to wear hearing protection when it is identified that their eight-hour time weighted average (TWA) is above the exposure action value of 90 dB. If subsequent monitoring shows that 85 dB is not surpassed for an eight-hour TWA, the employee is no longer required to wear hearing protection.

In the European Union, directive 2003/10/EC mandates that employers shall provide hearing protection at noise levels exceeding 80 dB(A), and that hearing protection is mandatory for noise levels exceeding 85 dB(A). Both values are based on 8 hours per day, with a 3 dB exchange rate. A 2017 Cochrane review found low-quality evidence that legislation to reduce noise in the workplace was successful in reducing exposure both immediately and long-term.

Sporting Events

Several sports stadiums pride themselves in having louder stadiums than their opponents because it may create a more difficult environment for opposing teams to play in. Currently, there are few studies on noise in sports stadiums, but some preliminary measurements show noise levels reaching 120 dB, and informal studies suggest that people may receive up to a 117% noise dose in one game. There are many challenges that face hearing conservationists such as sports culture. Sports fans will create noise in an attempt to distract other teams, and some sports teams have been known to create artificial noise in an attempt to make the stadium louder. In doing that, workers, teams, and fans may be at potential risk for damage to the auditory system.

NIOSH conducted a health hazard evaluation and studies at Monster Trucking and Stock Car racing events, spectators average noise levels ranged from 95 to 100 dBA at the Monster Truck event and over 100 dBA at the stock car racing event. NIOSH researchers also published noise exposure levels for drivers, crew members, and staff. Noise levels at the Bristol Motor Speedway ranged from 96 dBA in the stands to 114 dBA for a driver inside a car during practice. Peak noise levels in the Pit area reached or exceeded 130 dB SPL, a level often associated with human hearing threshold for pain. Several prominent NASCAR drivers have complete or partial hearing loss and other symptoms from their many years of exposure.

During the FIFA World Cup in 2010, noise levels created by fans blowing Vuvuzela averaged 131 dBA at the horn opening and 113 dBA at 2 meter distance. Peak levels reached 144 dB SPL, louder than a jet engine at takeoff.

A study of occupational and recreational noise exposure at indoor hockey arenas found noise levels from 81 dBA to 97 dBA, with peak sound pressure levels ranging from 105 dB SPLto 124 dB SPL. Another study examined the hearing threshold of hockey officials and found mean noise exposures of 93 dBA. Hearing threshold shifts were observed in 86% of the officials (25/29).

In a study of noise levels at 10 intercollegiate basketball games showed noise levels in 6 of the 10 basketball games to exceed the national workplace noise exposure standards, with participants showing temporary threshold levels at one of the games.

While there is no agency that currently monitors sports stadium noise exposure, organizations such as NIOSH or OSHA use occupational standards for industrial settings that some experts feel could be applied for those working at sporting events. Workers often will not exceed OSHA standards of 90 dBA, but NIOSH, whose focus is on best practice, has stricter standards which say that when exposed to noise at or exceeding 85 dBA workers need to be put on a hearing conservation program. Workers may also be at risk for overexposure because of impact noises that can cause instant damage. Experts are suggesting that sports complexes create hearing conservation programs for workers and warn fans of the potential damage that may occur with their hearing.

Studies are still being done on fan exposure, but some preliminary findings show that there are often noises that can be at or exceed 120 dB which, unprotected, can cause damage to the ears in seconds.

Mechanisms

NIHL occurs when too much sound intensity is transmitted into and through the auditory system. An acoustic signal from a sound source, such as a radio, enters into the external auditory canal (ear canal), and is funneled through to the tympanic membrane (eardrum), causing it to vibrate. The vibration of the tympanic membrane drives the middle ear ossicles, the malleus, incus, and stapes to vibrate in sync with the eardrum. The middle ear ossicles transfer mechanical energy to the cochlea by way of the stapes footplate hammering against the oval window of the cochlea, effectively amplifying the sound signal. This hammering causes the fluid within the cochlea (perilymph and endolymph) to be displaced. Displacement of the fluid causes movement of the hair cells (sensory cells in the cochlea) and an electrochemical signal to be sent from the auditory nerve (CN VIII) to the central auditory system within the brain. This is where sound is perceived. Different groups of hair cells are responsive to different frequencies. Hair cells at or near the base of the cochlea are most sensitive to higher frequency sounds while those at the apex are most sensitive to lower frequency sounds. There are two known biological mechanisms of NIHL from excessive sound intensity: damage to the structures called stereocilia that sit atop hair cells and respond to sound, and damage to the synapses that the auditory nerve makes with hair cells, also termed "hidden hearing loss".

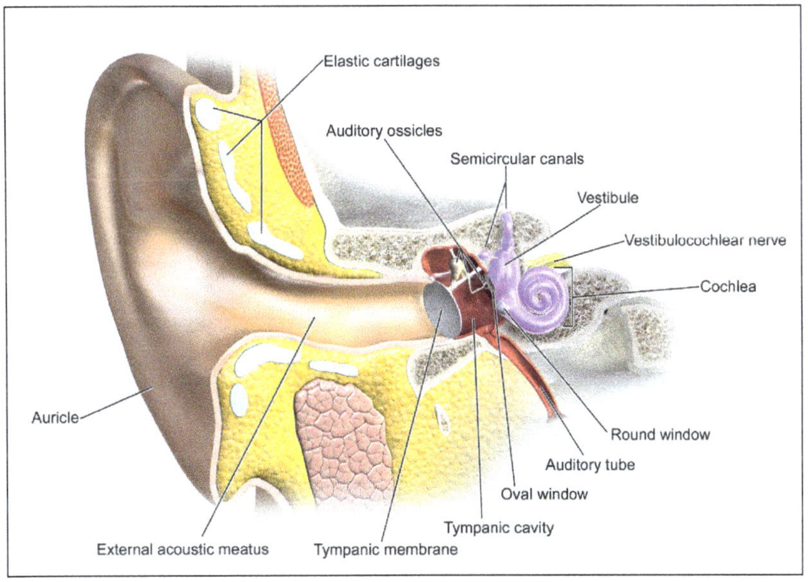

The anatomy of the ear.

Physiological Response

The symptoms are the external signs of the physiological response to cochlear overstimulation. Here are some elements of this response:

- Damaged sensory hairs (stereocilia) of the hair cells; damaged hair cells degenerate and die. In humans and other mammals, dead hair-cells are never replaced; the resulting hearing loss is permanent.
- Inflammation of the exposed areas. This inflammation causes poor blood flow in the exposed blood vessels (vascular stasis), and poor oxygen supply for the liquid inside the cochlea (endolymphatic hypoxia) Those noxious conditions worsen the damaged hair cell degeneration.
- Synaptic damages, by excitotoxicity. Noise over-stimulation causes an excessive release of glutamate, causing the postsynaptic bouton to swell and burst. However the neuron connection can be repaired, and the hearing loss only caused by the "wiring" (i.e. excitotoxicity) can thus be recovered within 2–3 days.

Hair Cell Damage or Death

When the ear is exposed to excessive sound levels or loud sounds over time, the overstimulation of the hair cells leads to heavy production of reactive oxygen species, leading to oxidative cell death. In animal experiments, antioxidant vitamins have been found to reduce hearing loss even when administered the day after noise exposure. They were not able to fully prevent it. Damage ranges from exhaustion of the hair (hearing) cells in the ear to loss of those cells. NIHL is therefore the consequence of overstimulation of the hair cells and supporting structures. Structural damage to hair cells (primarily the outer hair cells) will result in hearing loss that can be characterized by an attenuation and distortion of incoming auditory stimuli.

During hair cell death 'scars' develop, which prevent potassium rich fluid of the endolymph from mixing with the fluid on the basal domain. The potassium rich fluid is toxic to the neuronal endings and can damage hearing of the entire ear. If the endolymph fluid mixes with the fluid on the basal domain the neurons become depolarized, causing complete hearing loss. In addition to complete hearing loss, if the area is not sealed and leakage continues further tissue damage will occur. The 'scars' that form to replace the damaged hair cell are caused by supporting hair cells undergoing apoptosis and sealing the reticular lamina, which prevents fluid leakage. The cell death of two supporting hair cells rapidly expands their apical domain, which compresses the hair cell beneath its apical domain.

Nerve Damage

Recent studies have investigated additional mechanisms of NIHL involving delayed or disabled electrochemical transmission of nerve impulses from the hair cell to and along the auditory nerve. In cases of extreme acute acoustic trauma, a portion of the postsynaptic dendrite (where the hair cell transfers electrochemical signals to the auditory nerve) can rupture from overstimulation, temporarily stopping all transmission of auditory input to the auditory nerve. This is known as excitotoxicity. Usually, this sort of rupture heals within about five days, resulting in functional recovery of that synapse. While healing, an over-expression of glutamate receptors can result in temporary tinnitus, or ringing in the ears. Repeated ruptures at the same synapse may eventually fail to heal, leading to permanent hearing loss.

Prolonged exposure to high intensity noise has also been linked to the disruption of ribbon synapses located in the synaptic cleft between inner hair cells and spiral ganglion nerve fibers, leading to a disorder referred to as cochlear synaptopathy or hidden hearing loss. This disorder is cumulative and over time, leads to degeneration of the spiral ganglion cells of the inner ear and overall dysfunction in the neural transmission between auditory nerve fibers and the central auditory pathway. The most common symptom of cochlear synaptopathy is difficulty understanding speech, especially in the presence of competing noise. However, this type of hearing impairment is often undetectable by conventional pure tone audiometry, thus the name "hidden" hearing loss.

Acoustic over-exposure can also result in decreased myelination at specific points on the auditory nerve. Myelin, an insulating sheath surrounding nerve axons, expedites electrical impulses along nerves throughout the nervous system. Thinning of the myelin sheath on the auditory nerve significantly slows the transmission of electrical signals from hair cell to auditory cortex, reducing comprehension of auditory stimuli by delaying auditory perception, particularly in noisy environments.

Individual Susceptibility Towards Noise

There appear to be large differences in individual susceptibility to NIHL. The following factors have been implicated:

- Missing acoustic reflex.
- Previous sensorineural hearing loss.
- A bad general health state: bad cardiovascular function, insufficient intake of oxygen, a high platelet aggregation rate; and most importantly, a high viscosity of the blood.
- Cigarette smoking.
- Exposure to ototoxic chemicals (medication or environmental chemicals that can damage the ear), including certain solvents and heavy metals.
- Type 2 diabetes.

Diagnosis

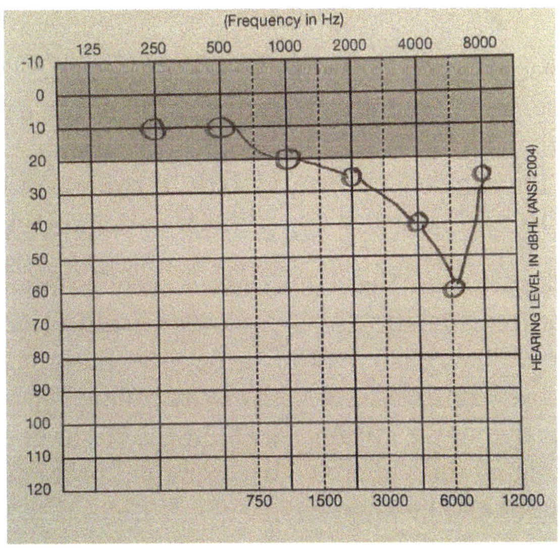

Example audiogram of a notch-shaped high frequency hearing loss.

Both NIHL caused by acoustic trauma and gradually-developed-NIHL can often be characterized by a specific pattern presented in audiological findings. NIHL is generally observed to decrease hearing sensitivity in the higher frequencies, also called an audiometric notch, especially at 4000 Hz, but sometimes at 3000 or 6000 Hz. The symptoms of NIHL are usually presented equally in both ears.

This typical 4000 Hz notch is due to the transfer function of the ear. Indeed, as any object facing a sound, the ear acts as a passive filter (although the inner ear is not an absolute passive filter, as the outer hair cells provide active mechanisms). A passive filter is a low pass : the high frequencies are more absorbed by the object, as high frequencies impose a higher pace of compression-decompression to the object. Thus, the high frequency harmonics of a sound are more harmful to the inner-ear.

However, not all audiological results from people with NIHL match this typical notch. Often a decline in hearing sensitivity will occur at frequencies other than at the typical 3000–6000 Hz range. Variations arise from differences in people's ear canal resonance, the frequency of the harmful acoustic signal, and the length of exposure. As harmful noise exposure continues, the commonly affected frequencies will broaden to lower frequencies and worsen in severity.

Prevention

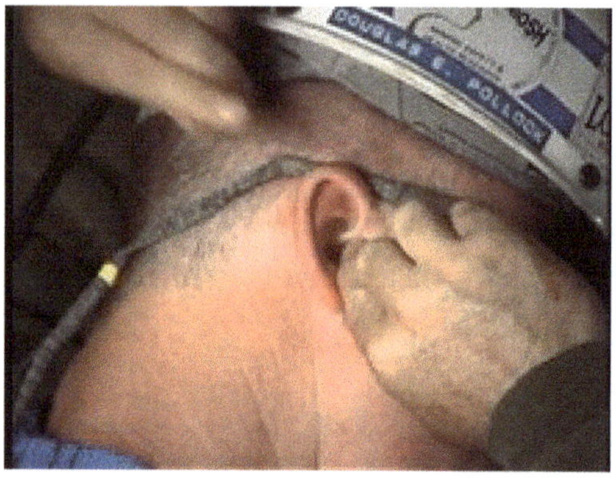

Usage of soft foam earplugs.

NIHL can be prevented through the use of simple, widely available, and economical tools. This includes but is not limited to personal noise reduction through the use of ear protection (i.e. earplugs and earmuffs), education, and hearing conservation programs. For the average person, there are three basic things that one can do to prevent NIHL: turn down the volume on devices, move away from the source of noise, and wear hearing protectors in loud environments.

Non-occupational noise exposure is not regulated or governed in the same manner as occupational noise exposure; therefore prevention efforts rely heavily on education awareness campaigns and public policy. The WHO cites that nearly half of those affected by hearing loss could have been prevented through primary prevention efforts such as: "reducing exposure (both occupational and recreational) to loud sounds by raising awareness about the risks; developing and enforcing relevant legislation; and encouraging individuals to use personal protective devices such as earplugs and noise-cancelling earphones and headphones."

Personal Noise Reduction Devices

Personal noise reduction devices can be passive, active or a combination. Passive ear protection includes earplugs or earmuffs which can block noise up to a specific frequency. Earplugs and earmuffs can provide the wearer with 10 dB to 40 dB of attenuation. However, use of earplugs is only effective if the users have been educated and use them properly; without proper use, protection falls far below manufacturer ratings. Higher consistency of performance has been found with custom-molded earplugs. Because of their ease of use without education, and ease of application or removal, earmuffs have more consistency with both compliance and noise attenuation. Active ear protection (electronic pass-through hearing protection devices or EPHPs) electronically filter out noises of specific frequencies or decibels while allowing the remaining noise to pass through. A personal attenuation rating can be objectively measured by using a hearing protection fit testing system.

Education

Education is key to prevention. Before hearing protective actions will take place, a person must understand they are at risk for NIHL and know their options for prevention. Hearing protection programs have been hindered by people not wearing the protection for various reasons, including the desire to converse, uncomfortable devices, lack of concern about the need for protection, and social pressure against wearing protection. Although youth are at risk for hearing loss, one study found that 96.3% of parents did not believe their adolescents were at risk, and only 69% had talked to their children about hearing protection; those aware of NIHL risks were more likely to talk to their teens.

A systematic review of the effectiveness of interventions to promote the use of hearing protection devices such as earplugs and earmuffs among workers found that tailored interventions improve the average use of such devices when compared with no intervention. Tailored interventions involve the use of communication or other types of interventions that are specific to an individual or a group and aim to change behavior. Mixed interventions such as mailings, distribution of hearing protection devices, noise assessments, and hearing testing are also more effective in improving the use of hearing protection devices compared with hearing testing alone. Programs that increased the proportion of workers wearing hearing protection equipment did reduce overall hearing loss.

Personal Listening Devices

Although research is limited, it suggests that increased exposure to loud noise through personal listening devices is a risk factor for noise induced hearing loss. More than half of people are exposed to sound through music exposure on personal devices greater than recommended levels. Research suggests stronger correlations between extended duration or elevated usage of personal listening devices and hearing loss.

Hearing Conservation Programs

Workers in general industry who are exposed to noise levels above 85 dBA are required by the Occupational Safety and Health Administration (OSHA) to be in a hearing conservation program (HCP), which includes noise measurement, noise control, periodic audiometric testing, hearing protection, worker education, and record keeping. Twenty-four states, Puerto Rico, and the U.S. Virgin Islands have OSHA-approved state plans and have adopted their own standards and enforcement policies. Most of these state standards are identical to those of federal OSHA. However, some states have adopted different standards or may have different enforcement policies. Most health and safety regulations are designed to keep damage risk within "acceptable limits" — that is, some people are likely to incur a hearing loss even when exposed to less than the maximum daily amount of noise specified in a regulation. Hearing conservation programs in other arenas (schools, military) have become more common, and it has been established that unsafe listening behaviors, such as listening to loud noise for extended periods of time without protection, persist despite knowledge of potential hearing loss effects.

However, it is understood that HCPs are designed to change behavior, which is known to be a complex issue that requires a multi-faceted approach. According to Keppler et al. in their 2015 study of such programming, they cite the necessary attitude change towards the susceptibility of risk and degree of severity of hearing loss. Among young adults, the concept of severity is most crucial because it has been found that behavior change may not occur unless an individual experiences NIHL or similarly related NIHL tinnitus, furthering warranting a multi-pronged approach based on hearing conservation programming and education.

Interventions to prevent noise-induced hearing loss often have many components. A 2017 Cochrane review found that hearing loss prevention programs suggest that stricter legislation might reduce noise levels. Giving workers information on their noise exposure levels by itself was not shown to decrease exposure to noise. Ear protection, if used correctly, has the potential to reduce noise to safer levels, but does not necessarily prevent hearing loss. External solutions such as proper maintenance of equipment can lead to noise reduction, but further study of this issue under real-life conditions is needed. Other possible solutions include improved enforcement of existing legislation and better implementation of well-designed prevention programmes, which have not yet been proven conclusively to be effective. The implications is that further research could affect conclusions reached.

Several hearing conservation programs have been developed to educate a variety of audiences about the dangers of NIHL and how to prevent it. Dangerous Decibels aims to significantly reduce the prevalence of noise induced hearing loss and tinnitus through exhibits, education and research. We're hEAR for You is a small non-profit that distributes information and ear plugs at concert and music festival venues. The Buy Quiet program was created to combat occupational noise exposures by promoting the purchase of quieter tools and equipment and encourage manufacturers to design quieter equipment. The National Institute on Deafness and Other Communication Disorders developed the It's a Noisy Planet. Protect their Hearing educational campaign to inform preteens, parents, and educators about the causes and prevention of NIHL. The National Institute for Occupational Safety and Health partnered with the National Hearing Conservation Association in 2007 to establish the Safe-in-Sound Excellence and Innovation in Hearing Loss Prevention Awards to recognize organizations that are successfully implementing hearing loss prevention concepts into their daily routines.

Medication

Medications are still being researched to determine if they can prevent NIHL. No medication has been proven to prevent or repair NIHL in humans.

There is evidence that hearing loss can be minimized by taking high doses of magnesium for a few days, starting as soon as possible after exposure to the loud noise. A magnesium-high diet also seems to be helpful as an NIHL-preventative if taken in advance of exposure to loud noises. Consuming sizable amounts of magnesium can be potentially harmful, so this treatment should be followed with caution.

Sound or Stress Training

Despite different people having different thresholds for what noises are painful, this pain threshold had no correlation with which noises cause hearing damage. The ear can not get more resistant to noise harmfulness by training it to noise. The cochlea is partially protected by the acoustic reflex, but being frequently exposed to noise does not lower the reflex threshold. It had been observed that noise conditioning (i.e. exposure to loud non-traumatizing noise) several hours prior to the exposure to traumatizing sound level, significantly reduced the damages inflicted to the hair-cells. The same "protective effect" was also observed with other stressors such as heat-shock conditioning and stress (by restraint) conditioning. This "protective effect" only happens if the traumatizing noise is presented within an optimum interval of time after the sound-conditioning session (-24 hours for a 15 min. sound-conditioning; no more protection after 48 hours). This "protective effect" had long been thought to involve the active mechanisms of the outer hair cells and the efferent system commanding them. The contractile effect of the outer hair cells, activated by the efferent nervous system has been proven to provide a protective effect against acoustic trauma.

However, a 2006 study revealed a different protective mechanism for stress conditioning. The study revealed that the stressor (sound, heat, or stress) conditioning increases the receptibility to glucocorticoid, a kind of anti-inflammatory hormone. The effects of glucocorticoid thus mitigate the inflammation from an acoustic trauma that can lead to hearing loss. In fact, high doses of corticoids are often prescribed by physicians after an acoustic-trauma in order to mitigate the inflammatory response.

Summarized, sound (or other stressor) conditioning is a pre-emptive medication against cochlea inflammation. It does not make the ear more resistant to noise. It reduces the inflammation caused by the acoustic trauma, which would cause subsequent damages to hair cells. While an anti-inflammatory medication would increase the quantity of anti-inflammatory hormone in the whole body, noise conditioning increases the number of receptors for the anti-inflammatory hormone, and only in the areas where it is much needed (i.e. cochlea).

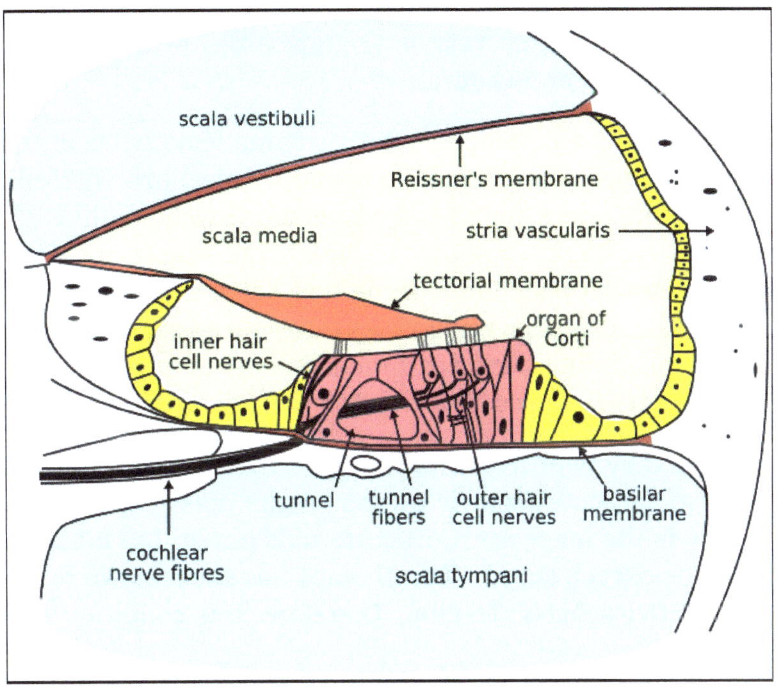

The inner hair cells are connected to afferent nerve fibers, and the outer hair cells are connected to efferent nerve fibers.

Physiological Response

- Stressor (noise, heat shock or stress) conditioning activates hormonal glands: the HPA axis. Note that the HPA axis is associated to the immune system.
- This HPA axis activation results in the up regulation of glucocorticoid receptors (GR) in the cochlea and the paraventricular nucleus (PVN) of the hypothalamus. Note that the glucocorticoid hormone is a kind of immune-reaction-inhibitor, including the inflammation reaction.

- This up regulation of GR thus prevents GR down regulation induced by acoustic trauma.
- The protective effect of noise-conditioning is blocked by adrenalectomy or pharmacological treatment with RU486 + metyrapone (a glucocorticoid receptor antagonist).

Treatment

Treatment options that offer "cures" for NIHL are under research and development. Currently there are no commonly used cures, but rather assistive devices and therapies to try and manage the symptoms of NIHL.

Acute Acoustic Trauma

Several clinical trials have been conducted to treat temporary NIHL occurring after a traumatic noise event, such as a gunshot or firework. In 2007, individuals with acute acoustic trauma after firecracker exposure were injected intratympanically with a cell permeable ligand, AM-111. The trial found AM-111 to have a therapeutic effect on at least two cases of those with acute trauma. Treatment with a combination of prednisolone and piracetam appeared to rescue patients with acute trauma after exposure to gunshots. However, those who received the treatment within an hour of exposure had higher rates of recovery and significantly lower threshold shifts compared to those who received treatment after one hour.

Additionally, clinical trials using antioxidants after a traumatic noise event to reduce reactive oxygen species have displayed promising results. Antibiotic injections with allopurinol, lazaroids, α-D-tocopherol, and mannitol were found to reduce the threshold shift after noise exposure. Another antioxidant, Ebselen, has been shown to have promising results for both TTS and PTS. Ebselen mimics gluthathione peroxide, an enzyme that has many functions, including scavenging hydrogen peroxide and reactive oxygen species. After noise exposure, gluthathione peroxide decreases in the ear. An oral administration of ebselen in both preclinical tests on guinea pigs and human trials indicate that noise induced TTS and PTS was reduced.

Recently, combination therapy with hyperbaric oxygen therapy (HBO) and corticosteroids has been found to be effective for acute acoustic trauma. Acute noise exposure causes inflammation and lower oxygen supply in the inner ear. Corticosteroids hinder the inflammatory reaction and HBO provides an adequate oxygen supply. This therapy has been shown to be effective when initiated within three days after acoustic trauma. Therefore, this condition is considered an ENT emergency.

Gradually Occurring NIHL

At the present time, no established clinical treatments exist to reverse the effects of permanent NIHL. However, current research for the possible use of drug and genetic therapies look hopeful. In addition, management options such as hearing aids and counseling exist.

Many studies have been conducted looking at regeneration of hair cells in the inner ear. While hair cells are generally not replaced through cell regeneration, mechanisms are being studied to induce replacement of these important cells. One study involves the replacement of damaged hair cells with regenerated cells, via the mechanism of gene transfer of atonal gene Math1 to pluripotent stem cells within the inner ear. Other atonal genes are being studied to induce regeneration of hair cells in the inner ear.

Management

For people living with NIHL, there are several management options that can improve the ability to communicate. These options include counseling, amplification, and other assisted listening devices, such as frequency modulation (FM) systems. FM systems can enhance the use of hearing aids and overcome the effects of poor listening conditions because the signal is sent from the microphone worn by the speaker directly to the listener. The prognosis has improved with the recent advancements in digital hearing aid technology, such as directional microphones, open-fit hearing aids, and more advanced algorithms. Hearing aids can mask or cover up the tinnitus, and many with hearing loss and tinnitus find relief by using hearing aids. Though there is no cure or agreed-upon treatment for tinnitus, some drugs have been shown to provide temporary reduction of tinnitus. Other treatments for tinnitus include cognitive-behavioral therapy, biofeedback, and electrical stimulation. Annual audiological evaluations are recommended to monitor any changes in a patient's hearing and to modify hearing-aid prescriptions.

A systematic-review conducted by the American Academy of Audiology Task Force On the Health-Related Quality of Life Benefits of Amplification in Adults found the use of hearing aids to increase quality of life. The review pertained to adults who experienced sensorineural hearing loss, which can be caused by excessive, loud noise.

Noise Control

Some of the ways to control noise pollution are as follows: (1) Control at Receiver's End (2) Suppression of Noise at Source (3) Acoustic Zoning (4) Sound Insulation at Construction Stages (5) Planting of Trees (6) Legislative Measures.

It is evident that noise is not merely a nuisance but is a serious environmental problem and a health hazard.

Like all other pollutions, noise pollution needs to be controlled by measures which will maintain the acceptable levels of noise pollution for human beings and buildings as indicated.

Noise pollution can be effectively controlled by taking the following measures:

Control at Receiver's End

For people working in noisy installations, ear-protection aids like ear-plugs, ear-muffs, noise helmets, headphones etc. must be provided to reduce occupational exposure.

Suppression of Noise at Source

This is possible if working methods are improved by:

- Designing, fabricating and using quieter machines to replace the noisy ones.

- Proper lubrication and better maintenance of machines.
- Installing noisy machines in sound proof chambers.
- Covering noise-producing machine parts with sound-absorbing materials to check noise production.
- Reducing the noise produced from a vibrating machine by vibration damping i.e. making a layer of damping material (rubber, neoprene, cork or plastic) beneath the machine.
- Using silencers to control noise from automobiles, ducts, exhausts etc. and convey systems with ends opening into the atmosphere.
- Using glass wool or mineral wool covered with a sheet of perforated metal for the purpose of mechanical protection.

Acoustic Zoning

Increased distance between source and receiver by zoning of noisy industrial areas, bus terminals and railway stations, aerodromes etc. away from the residential areas would go a long way in minimising noise pollution. There should be silence zones near the residential areas, educational institutions and above all, near hospitals.

Sound Insulation at Construction Stages

- Sound travels through the cracks that get left between the door and the wall. For reducing noise, this space (jamb frame gap) should be packed with sound absorbing material.
- Sound insulation can be done by constructing windows with double or triple panes of glass and filling the gaps with sound absorbing materials.
- Acoustical tiles, hair felt, perforated plywood etc. can be fixed on walls, ceilings, floors etc. to reduce noise (especially for sound proof recording rooms etc.)

Planting of Trees

Planting green trees and shrubs along roads, hospitals, educational institutions etc. help in noise reduction to a considerable extent.

Legislative Measures

Strict legislative measures need to be enforced to curb the menace of noise pollution. Some of these measures could be:

- Minimum use of loudspeakers and amplifiers especially near silence zones.
- Banning pressure horns in automobiles.
- Framing a separate Noise Pollution Act.

Noise Regulation

Noise regulation includes statutes or guidelines relating to sound transmission established by national, state or provincial and municipal levels of government. After the watershed passage of the United States Noise Control Act of 1972, other local and state governments passed further regulations.

A noise regulation restricts the amount of noise, the duration of noise and the source of noise. It usually places restrictions for certain times of the day.

Although the United Kingdom and Japan enacted national laws in 1960 and 1967 respectively, these laws were not at all comprehensive or fully enforceable as to address generally rising ambient noise, enforceable numerical source limits on aircraft and motor vehicles or comprehensive directives to local government.

Local Noise Ordinances in U.S. and Europe

Local ordinances are principally aimed at construction noise, power equipment operated by individuals and unmuffled industrial noise penetrating residential areas. Thousands of U.S. cities have prepared noise ordinances that give noise control officers and police the power to investigate noise complaints and enforcement power to abate the offending noise source, through shutdowns and fines. In the 1970s and early 1980s there was even a professional association for noise enforcement officers called NANCO, "National Association of Noise Control Officials".

Today only a handful of properly trained Noise Control Officers remain in the United States. A typical noise ordinance sets forth clear definitions of acoustic nomenclature and defines categories of noise generation; then numerical standards are established, so that enforcement personnel can take the necessary steps of warnings, fines or other municipal police power to rectify unacceptable noise generation. Ordinances have achieved certain successes but they can be thorny to implement. Many European cities are still treating noise as the U.S. did in the 1960s, as a nuisance and not as a numerical standard to be achieved.

Effects of Noise on Health and Welfare

One obligation of a community is to protect its citizens from adverse environmental influences. Noise is one of these factors, Noise has documented effects on people, they can be divided into three types. The first type is a physical effect that directly and adversely effects a person's health. Hearing loss and vibration of bodily components are examples. The second type is a physiological effect that adversely effects a person's health; heightened blood pressure and general stress response are examples. The third type is psychological that adversely effects a person's welfare; examples are distraction, annoyance, and complaint. The only feasible legal basis for a community's right to control noise is based on these adverse health and welfare effects. It is clearly easier to uphold the constitutionality of a noise ordinance in a court of law if it can be shown that it is based on health and welfare concerns. The following is a short list of recognized effects of noise that can be addressed as a reason for a noise ordinance.

Excess non-Occupational noise exposure, hearing loss on both public and private property, speech interference on both public and private property, audio interference on both public and private property, and sleep interference on mostly private property.

Some Legal Considerations in the United States

There are several fundamental issues that shape the legality, effectiveness and enforceability of any community noise regulation.

Preemption

The federal government has preempted certain areas of noise regulation. They can be found in the Code of Federal Regulations under the EPA Noise Abatement Programs; Parts 201 to 205 and 211 cover railroads, motor carriers in interstate commerce, construction equipment, motor vehicles. They require product labeling and prohibit tampering with noise control devices. Communities may enact regulations that are no more strict than the federal ones so that local enforcement can be carried out. They can enact curfews and restrict vehicle use in established zones such as residential. Any restriction on interstate motor carriers or railroads may NOT be for the purpose of noise control. States have police powers granted by the Constitution. They may also enact regulations that are no more strict than federal regulations. They may also preempt local ordinances. California and New Jersey have comprehensive noise codes that communities must meet. Many states required that local ordinances be no more strict than the state code whether such code exists or not. One relatively common preemption is protection of shooting ranges from noise regulation or litigation and right to farm laws that protect agricultural areas from nuisance litigation by encroaching residential areas.

Constitutional Vagueness

In one case, the United States Supreme Court has said Void for vagueness "It is a basic principle of due process (Amendment V) that an enactment is void for vagueness if its prohibitions are not clearly defined. Vague laws offend several important values. First, because we assume that man is free to steer between lawful and unlawful conduct, we insist that laws give the person of ordinary intelligence a reasonable opportunity to know what is prohibited, so that he may act accordingly. Vague laws may trap the innocent by not providing fair warning. Second, if arbitrary and discriminatory enforcement is to be prevented, laws must provide explicit standards for those who apply them. A vague law impermissibly delegates basic policy matters to policemen, judges, and juries for resolution on an ad hoc and subjective basis, with the attendant dangers of arbitrary and discriminatory application. Third, but related, where a vague statute abuts upon sensitive areas of basic First Amendment freedoms, it operates to inhibit the exercise of those freedoms. (the chilling effect). In one case, the court declared that numerical sound levels were constitutional as well as the term plainly audible provided it was associated with a reasonable distance. Two requirements for a noise ordinance provision is that:

- Provide fair warning.
- Avoid the possibility of arbitrary enforcement.

Free Speech

Examples are: slander, hate speech, and falsely calling "fire" in an assembly. Generally, the time, place, or manner of restrictions must:

- Be content neutral.
- Be narrowly tailored.
- Serve a significant government interest.
- Leave open alternative channels of communication.

One example concerned protesters outside an abortion clinic. The US Supreme Court stated that limitations placed on noise making were necessary to ensure the well-being of patients.

Overbreadth

The US Supreme Court has addressed the issue of too much regulation by defining overbreadth as: A clear, precise enactment may nevertheless be overbroad if, in its reach, it prohibits constitutionally protected conduct. One means to avoid being overbroad is by having provisions that are specific so that potential violators are given fair warning of what is prohibited. In one case the court ruled that the specificity of the city ordinance regulating school verbal protests was not constitutionally vague, gave fair warning, and was not an invitation to arbitrary enforcement and so was not overbroad, despite the implied limitation on free speech.

Nuisance

Nuisances are part of tort law. Torts are civil wrongs whether intentional or accidental from which injury occurs to another. Nuisances can be public or private. A public nuisance is an unreasonable interference with a right common to the general public while a private nuisance is an interference with a person's enjoyment and use of his land. Black's Law Dictionary defines a nuisance as. That which annoys and disturbs one in possession of his property, rendering its ordinary use and occupation physically uncomfortable to him. It defines a private nuisance as anything done to the hurt or annoyance of the lands, tenements, or hereditaments of another. As distinguished from public nuisance, it includes any wrongful act which destroys or deteriorates the property of an individual or of a few persons or interferes with their lawful use or enjoyment thereof, or any act which unlawfully hinders them in the enjoyment of a common or public right and causes them a special injury different from that sustained by the general public. The word nuisance is connected closely with the word disturbance which is used extensively in many noise ordinances. A definition below further restricts the meaning to noise disturbance. Nuisance law applies to both community noise regulation as well as private suits brought to court to reduce noise impact.

Enforceability

Care must be taken in writing a subjective noise provision so that it overcomes the objections listed above. Care must be taken when writing an objective noise provision to make sure that the sound levels are physically realizable. For example, requiring the maximum sound level of an automobile to be 40 dB(A) or the maximum sound level in a residential zone to be 30 dB(A) opens the provision to an enforceability challenge.

Confiscation

Confiscation is the taking of private property for public use without compensation. It may occur legally when the government seizes property used in illegal practices. Confiscation may occur without an arrest of a person because it is seen as an arrest of property, rather than a person, and the necessity of finding a party guilty is not required in all cases. In many cases, the guilt or innocence of the property owner is irrelevant and the government need not prove anything beyond a reasonable doubt. In order to seize property, there must be either a warrant or a showing of probable cause. The mere showing of probable cause to support the forfeiture of an individual's property has been a controversial issue. Court opinion varies, but probable cause most often means that the government need only demonstrate reasonable grounds for the belief that the property is properly subject to forfeiture, which can be supported by something less than prima facie proof, but more than mere suspicion. The Peoria, Illinois noise ordinance allows confiscation of vehicles that emit excessive sound caused by sound producing devices. Colorado Springs, Colorado permits confiscation of sound producing devices.

Four Types of Noise Regulation

Fixed sound sources must be treated differently than moving sources. In the former case, the listener is normally defined while for moving sources it is not. Historically, regulations were enforced by the subjective judgment of an enforcing officer. With the advent of sound measuring equipment, the judgment can be based on measured sound levels. Most comprehensive noise ordinances contain four types of provisions.

Subjective Emission These regulations allow an official to decide if the output of a sound source is acceptable without recourse to sound measurements and without regard to the presence of a specific listener. Regulations with plainly audible terms on public property as a criterion are examples.

Subjective Immission These regulations allow an official to decide if the sound received by a listener is acceptable without recourse to sound measurements and without regard for the specific sound power generated by the source. Regulations with plainly audible or noise disturbance terms on private property as a criterion are examples.

Objective Emission These regulations require an official to measure the output of a sound source to determine whether it is acceptable without regard to the presence of a specific listener. Regulations with specific maximum sound output levels for motor vehicles are examples.

Objective Immission These regulations require an official to measure the sound received by a listener to determine whether it is acceptable without regard to the specific sound power generated by the source. Regulations with maximum allowable sound levels on property lines are examples.

Some Definitions used in the United States

Many communities have definitions that are local to them, such as those defining motor vehicles and sound levels and sound level measurements. Some that have been added to make noise enforcement more specific are listed here.

Engine Braking Device

A compression braking device installed on large motor vehicles to assist in reduction, or control, of vehicle speed. When activated, the engine converts from a power source to a power absorber by converting the engine into an air compressor.

Muffler

Any device for the abatement of sound emission while permitting the transfer of gas. A muffler is considered to be in good working order if the sound reduction is equal to, or greater than, that of the original equipment.

Noise Disturbance

Any sound or vibration which:

- May disturb or annoy reasonable persons of normal sensitivities or;
- Causes, or tends to cause, an adverse effect on the public health and welfare or;
- Endangers or injures people or;
- Endangers or injures personal or real property.

This can also be defined as noise nuisance.

Place of Public Entertainment

Any location, exterior, or interior, to a building that regularly permits public entrance for entertainment purposes. For this purpose, "public" means citizens of all types, including but not limited to, children, and private or public employees.

Plainly Audible Sound

Any sound for which the information content is unambiguously communicated to the listener, such as, but not limited to, understandable speech, comprehension of whether a voice is raised or normal, repetitive bass sounds, or comprehension of musical rhythms, without the aid of any listening device.

Powered Model Vehicle

Any self-propelled airborne, water-borne, or land-borne, plane, vessel, or vehicle, which is not designed to carry persons, including, but not limited to, any model airplane, boat, car, or rocket.

Real Property Boundary

An imaginary line along the ground surface, and its vertical extension, which separates the real property owned by one person from that owned by another person, but not including intra-building real property divisions.

Sound Reproduction Device

Any device, instrument, mechanism, equipment or apparatus for the amplification of any sounds from any radio, phonograph, stereo, tape player, musical instrument, television, loudspeaker or other sound-making or sound-producing device or any device or apparatus for the reproduction or amplification of the human voice or other sound.

Vibration Perception Threshold

The minimum ground or structure borne vibrational motion necessary to cause a person of normal sensitivity to be aware of the motion through contact, hearing, or through visual observation of moving things.

Ordinance Provisions for Stationary Sources in the United States

There are three levels of regulation for stationary sound sources. The most basic is the general one associated with noise disturbance. It is a very broad subjective immission control that has evolved from earlier disturbance of the peace provisions. Subjectivity can lead to arbitrary enforcement. The next level of regulation is less broad; it is an objective immission control that uses specific levels of sound considered to be a noise disturbance. Arbitrary enforcement is reduced. In both cases, however, the person creating the sound may not be aware that his actions are in violation. The concept that a potential violator should have fair warning that his actions are in violation has led to provisions that address specific noise problems. The sections below list those that are found in community noise ordinances.

Air Conditioning, Heating and Pool Equipment

This provision is a subjective immission control. An evaluation of the noise disturbance is made at the listener without a sound level meter. This provision is mostly applied in residential zones such as homes, apartments and condominiums. Albuquerque, NM requires that such units the maximum permissible sound levels and recommend that those units be placed away from other residential units or on roof tops to diminish impact.

Airport and Airport Operations

Community control of airport created noise is limited to those sounds not related to flight operations. The community is able to control the land use around the airport however.

Animals

This provision is a subjective immission control. Most relate to barking dogs and put an upper time limit for continuous sound from them. New Jersey considers a violation if the sound is continuous for more than 5 minutes or intermittent for more than 20 minutes. They also consider it a defense to violation if the animal is provoked to bark. Connecticut exempts animal sounds while Anchorage, AK requires that continual violations permit the animal to be taken and put out for adoption.

Authorized Outdoor Discharge of Firearms and Shooting Ranges

This provision contains only a curfew since most states protect shooting ranges from liability for noise disturbance. It can include a curfew requirement and a requirement for a public hearing if expansion of the range is desired. The provision may prohibit other weapons such as rocket propelled projectiles but may exempt unpowered weapons such as arrows. South Carolina requires that a sign stating shooting range-noise area be placed on all primary roads. Arizona (ARS 17-602) places a curfew from 10pm to 7 am. It also allows a tradeoff between the number of events and the maximum permitted sound level. New York also trades off overall levels with the duration of the sounds. Colorado declared that noise restrictions on shooting ranges is a detriment to public health, welfare, and morale.

Condominiums and Apartments

This provision is a subjective immission control. It is designed to limit the noise disturbance between living units as defined by an enforcement official. One criterion used to evaluate that disturbance is use of plainly audible but at the location of the listener instead of at a specific distance. However, Charlotte, NC limits indoor levels to 55 dB(A) be tween 9am to 9pm and 5 dB less at other times, but only from sound reproducing devices. Salt Lake Valley Health Department, Minneapolis, MN, and Albuquerque, NM use levels to existing ambient to define a violation. Albuquerque NM and Omaha, NE require that intruding sounds not be audible. Burlington, VT requires that renters be supplied with the city noise ordinance.

Construction

This provision can be both a subjective immission control and an objective emission control. Normally there are daily curfews and in some case weekend curfews. The subjective aspect is to prevent noise disturbance in the adjacent community. The objective aspect is to control the sound output of specific machines. There are four major sources of site noise: (1) direct sound from continually operating equipment such as air compressors; (2) intermittent sound from equipment such as jack hammers;(3)backup alarms; and (4) hauling equipment such as trucks. Air compressor noise is regulated by CFR 204 and backup alarms are regulated by CFR 1926. Boston, MA permits construction on weekdays between 7am and 6 pm. Madison, WI limits sound levels to 88 dB(A) at 50 Feet. Miami, FL considers the noise a noise disturbance if it occurs between 6pm and 8am during the week and any time on Sunday. Dallas, TX permits construction in residential zones from 7am to 7pm on weekdays, from 9am to 7pm on Saturdays and Holidays, and prohibits construction on Sundays. Albuquerque, New Mexico has a more complex control. It prohibits construction and demolition within 500 feet of noise sensitive properties (residences included) if the equipment sound control devices are less effective than the original equipment and if noise mitigation measures are not used when the levels exceed 90 dB (weighting not specified) or more than 80 dB during the day for three days.

Domestic and Commercial Power Tools

This provision is a subjective immission control with a curfew. It is used in residential zones as well as in commercial areas abutting residential zones. Portland, OR has several ways to handle these tools. They separate outdoor and indoor use to different maximum levels at the property line. They have a night curfew and separate 5 HP tools from higher powered tools. Madison WI has similar HP restrictions. Albuquerque, NM restricts locations to more than 500 feet from residential and noise sensitive zones. Dallas, TXexempts lawn maintenance tools during daylight hours. Green Bay, WI exempts snow removal tools.

Explosives, Firearms, Impulsive Sources and Similar Devices

This provision is a subjective immission control with a curfew. It is for impulsive sound sources that are not associated with construction activities or shooting ranges. Many communities use the Maximum Permissible Sound Levels criterion, with a correction for the character of the sound. Illinois () sets maximum blasting levels by land use zone and in three time categories. Portland, OR limits levels to 100 dB (peak) from 7am to 10pm and 80 dB (peak) from 10pm to 7am.

Fracking Operations

Hydraulic Fracturing operations generate site sound as well as vehicle sound and several different provisions are required to control it. Federal law regulates the levels of certain site machinery. A subjective immission control or an objective immission control can be applied to surrounding neighborhoods. Motor vehicle sound is mostly off-site so vehicle noise regulations are applicable. The State of New York has announced a statewide ban on such operations. Buffalo, NY and Pittsburgh, PA have announced a ban. Colorado has numerous activities to stop fracking.

Funerals

This is a subjective emission control to reduce the excessive shouting and protests that can surround funeral proceedings. It makes use of the plainly audible term and so adds a distance criterion. There are certain groups, particularly those that object to involvement in foreign wars, who believe it is an obligation to disrupt and picket funerals, especially those of deceased military veterans. This provision must not infringe on Constitutionally protected free speech. Illinois h (720 ILCS 5/26 6) has a comprehensive provision covering more aspects of this event than noise. They failed to use “plainly audible” which is narrower than “audible”. Utah restricts disruptive activity to beyond 200 feet.

Loading and Unloading Operations

This provision is a subjective immission control with a curfew. Operations in commercial facilities can impact adjacent residential zones. Los Angeles, CA places a curfew on such operations between 10pm and 7am but only if the source is within 200 feet of a residence. Chicago, IL permits night operations unless they create a noise disturbance. Hammond, IN prohibits noise disturbance between 7pm and 7am.

Maximum Permissible Sound Pressure Levels

This provision is an objective immission control. It requires the measurement of sound levels at or beyond a property line and its vertical extension. There are several methods for implementing such a provision:

- It may not permit any exceedence or may permit exceedence only for a percentage of the measurement period.

- It may require the measurement method to be instantaneous, such as dB(A) or time-averaged, such as Energy Equivalent Level (Leq).
- It may be a fixed level limit, such as 55 dB(A), or it may be a level relative to the ambient sound, such as 5 dB(A) above the ambient.
- It may require measurement of the frequency spectrum, such as one octave bands, or A-weighting, such as dB(A).
- It may define different maximum levels based on zoning criteria, such as residential, commercial, or industrial.
- It may define different maximum levels based on time-of-day or day-of-week, such as reduced maxima during night hours or on weekends.
- It may require reduction of maximum levels based on the character of the sound, such as intermittent or impulsive.
- It may exempt certain classes of sound sources, such as shooting ranges, farm equipment, emergency equipment, railroads, or licensed activities.

Most noise ordinances set maximum levels for two time periods: Day (7am to 10pm) and Night (10pm to 7am). San Diego sets three periods: Day (7am to 7pm), Evening (7pm to 10pm), and Night (10pm to 7am) and exempts industrial zones from time based restrictions. Seattle, WA sets two time periods but changes 7am to 9am on weekends and holidays. Several states have maximum permissible land use sound levels in dB(A). Most have Day and Night periods and three use categories: residential, Commercial and Industrial. Washington sets maximum levels in dB(A) but allows 5 dB(A) more if the sound is only 15 minutes in an hour, or 10 dB(A) for 5 minutes in an hour. Numerous cities have fixed levels, permitting excess levels for short times, while others use L_{eq}. Los Angeles, CA uses a relative level with a stated but presumed ambient. New York City, NY requires L_{eq} measurements to be made over one hour. Atlanta, GA limits impulsive sound to 100 dB(C) at property lines, while most reduce the maximum level by 5 dB for pure tones and impulsive sounds.

Motor Vehicle or Motorboat Repairs and Testing

This provision is a subjective immission control with a curfew. If the activity is done in a residential zone, the Domestic Tools provision can be applied for the repairs, but this provision also is used for the testing phase of any repairs. Los Angeles, CA covers this violation in three ways. The first is application of the noise disturbance provision in residential districts between the hours of 8 pm to 8 am. The second is being plainly audible at a distance of 150 feet or more in residential districts between the hours of 8 pm to 8 am. The third is exceeding the presumed ambient by 5 dB. Hammond, IN prohibits this activity as a noise disturbance at any time.

Noise Disturbance Prohibited

This provision is a subjective immission control. This provision is generic in that it covers all events that are considered disturbing by a listener, with or without measurement. The strength of this provision is that it covers situations not contemplated in a noise ordinance and can be used as backup for more specific provisions. The weakness is that it may not give fair warning, may lead to arbitrary enforcement on the part of the regulator, or permit unreasonable demands by a listener.

Noise Sensitive Zones

This provision is a subjective immission control. It is used to reduce levels of both stationary and vehicular sound sources around hospitals, schools, and other noise sensitive locations. To provide fair warning visible signs must be posted. It is possible to have an extensive list. For example, if churches are on the list and the community has many of them, signage, compliance, and enforcement can be a problem. In modern hospital environments, helicopter sound is exempt.

Places of Public Entertainment

This provision is an objective immission control. It regulates the site of the sound source while the Sound Reproduction Devices section regulates the devices that create the sound. It can regulate the sound levels received by involuntary listeners in the surrounding community as well as the sound levels received by voluntary listeners. If the latter aspect is incorporated, limiting internal sound levels often resolves community noise impact. Los Angeles requires warning signs and limits noise exposure to 95 dB(A) at any position normally occupied. Seattle, WA considers the sound emitted to be in violation if the sound is plainly audible within a dwelling from 10 pm to 7am; the need for a sound level meter is avoided. Chicago, IL limits received sound levels to 55 dB(A) inside a residential dwelling unit but if the ambient is greater, the limit is 65 dB(A). If outdoors, the limit is conversational level at 100 feet from the property line. If the building is set back 20 feet from the property line, the allowable level is 84 dB(A). Both of Chicago's limits apply from 10 pm to 8 am. Salt Lake health Department, UT sets the limit at 95 dB(A) at a position that would normally be occupied by a patron and 100 dB(A) at other positions. They also require a sign stating warning: sound levels on this premise [sic] may cause permanent hearing damage. hearing protection is available. Anchorage, AK sets maximum levels for any patron at 90 dB(A).

Powered Unmanned Vehicles or Engines

This provision is a subjective immission control with a curfew. It has been used to regulate the sound of model aircraft on both private and public property. It applies to airborne, water-borne, and land-borne, unmanned vehicles. It makes no distinction between model vehicles and full-size unmanned vehicles. it applies also to the engines of those vehicles. Most regulations pertained to private unmanned vehicles, normally restricted to local open fields. The development of drones with microphones, cameras and GPS has opened the door to commercial use over wider private and public properties. Since federal preemption of drone use will likely occur, it is important for this provision to make the distinction. Salt Lake Health Department limits activity to 800 feet from a dwelling between 10 pm and 7 am, or if it cases a noise disturbance. Atlanta, GA uses the plainly audible criterion across a residential property line, on a public property from 10 pm to 7 am on weekdays or from 10 pm to 10 am on weekends or holidays for any sound source.

Propane Cannons

This provision is a subjective immission control with a curfew. A propane cannon is used to keep animals and birds from destroying commercial crops. In large fields, many are used and fired as frequent intervals. The sound levels are equivalent to the firing of a small artillery cannon. The provision may contain requirements that limit the number of cannons permitted in a specific area and the number firings per hour for each cannon. Many US states have a Right to Farm Act that limits nuisance litigation. Florida stated that it was a purpose of their act to protect reasonable agricultural activities conducted on farm land from nuisance suits. They also added a section that limited expand of operations without consideration of noise. Fairfax County, VA require agricultural operation to meet maximum land use regulations and prohibit unnecessary noise. British Columbia Ministry of Agriculture have developed a comprehensive set of rules for cannon use.

Public Address Systems

This provision is a subjective immission control. It can contain a plainly audible term or a curfew. It is applied to commercial facilities using a sound system to deliberately propagate mostly speech, but also music. Most cities have provisions relevant to this subject. Lakewood, CO used plainly audible as a regulatory tool and prohibits the sounding of bells, or chimes from 10pm to 7am. Charlotte, NC limits levels to 60 dB(A) at 50 feet from 9am to 9pm and 50 dB(A) at other times. Indianapolis, IN addresses broadcasts from aircraft. Connecticut exempts bells, carillons, and chimes from religious facilities.

Sound Reproduction Devices

This provision is subjective immission control. It may contain a numerical level or a plainly audible term and a curfew. it is applied to specific sources of sound as opposed to any location at which the sound is created. It is applied primarily to amplified sound sources. Older provisions listed several items such as televisions, phonographs, etc. Changing the title to the above addresses the real issue and allows for novel sound production devices. Numerous communities have provisions for these devices; many use plainly audible as the criterion, such as Omaha, NE and Buffalo NY.

Stadiums and Outdoor Music Festivals

This provision can be either a subjective or objective immission control with a curfew. The subjective aspect relates to noise disturbance in the local community. The objective aspect limits the acceptable sound level in the local community. Illinois exempts certain stadiums and exempts festivals, parades, or street fairs. Colorado Spring, CO has similar exemptions, but limits the sound levels to 80 dB(A) at residential locations.

Stationary Emergency Signaling Devices

This provision is an emission control with a list of devices that are exempt. It can have a term that limits the time periods in which emergency alarms may be tested. It can have a term that limits the activation time of burglar or fire alarms. Chicago, IL limits the time for tests to 4 minutes between 9am and 5pm. Oregon prohibits sound when an emergency vehicle is stationary.

Stationary Non-emergency Non-safety Signaling Devices

This provision is an emission control that limits the activation period of alarms and restricts activation to a specific time-of-day or day-of-week. Los Angeles, CA prohibits the sounding if the signal can be heard at 200 feet or more. Chicago, IL considers the sound to be a noise disturbance in residential areas if the sound exceeds 5 minutes in any hour; steam whistles are exempt. Albuquerque, NM restricts levels to 5 dB over the ambient at a property line and applies maximum permissible residential level as well as plainly audible restrictions at night.

Street Sales

This provision is a subjective immission control with a curfew. Boston, MA prohibits street sales near schools or churches if there is a "disturbance of the peace". Hammond, IN places a curfew between 6 pm and 9 am.

Tampering

This provision prevents the modification of muffling devices that increase the emitted sound. It can also be used to prevent the commercial sale of such mufflers. Most states and communities have prohibitions on tampering with noise reduction devices. whether stationary or moving. Salt Lake County health Department prohibits modifications of mufflers that increase sound levels and prohibits tampering with noise rating labels.

Vibration

This provision is a subjective emission control. Noise disturbance caused by vibration comes in three forms. One is contact with vibrating surfaces, the second in auditory, and the third is the observation of surrounding objects movement. Objectively regulating vibration is difficult so this provision makes use of the Vibration Perception Threshold. Railroad caused vibration is preempted by federal law (CFR 201). Chicago, IL uses the perception threshold method. Dallas, TX requires measurement of low frequency vibration. Maryland uses the definition of noise to include sound and vibration at sub-audible frequencies.

Wind Turbines

The sound created by wind turbines is caused by the blade rotation similar to aircraft propellers. Because the rotation rate is low, the frequency is also low, but the large size of many can result in disturbing sound levels, particularly in high wind areas. Most local control is done by advantageous site planning. New Hampshire sets a sound level limit of 55 dB when measured at the site property line, allowing for exceptional events, such as storms.

Ordinance Provisions for Moving Vehicles in the United States

Stationary sources have fixed positions, so it is possible to define the listeners and therefore immission controls are appropriate. Motor vehicles are moving sources so it is not possible to define any specific listeners so emission controls are appropriate. There are exceptions to this distinction. Construction equipment, and some recreation vehicles, move within a bounded area and can be considered to be time varying fixed sources. Standing motor vehicles can radiate sufficient sound to create noise disturbance. These must be treated by specific provisions.

Adequate Mufflers or Sound Dissipation Devices

This provision is an objective emission control. Unlike the Tampering provision, this is specific to motor vehicles. It requires that a vehicle muffler not create more sound than the original equipment which has been measured. It prohibits any modification or replacement that increases the sound emission beyond that of the original equipment. It prohibits the sale of mufflers that do not meet original equipment standards. Many states have requirements that a muffler shall be in good working order which is not specific enough. California requires that a retail seller that sells a product in violation of the muffler regulation must install a replacement muffler that meets the regulation and must reimburse the purchaser for the expense of replacement.

Airboats and Hovercraft

This provision is both an objective emission control and a subjective immission control. It can set a maximum sound level at a specified distance (typically 50 feet). It can have a curfew. It can be based on noise disturbance in the community. It can also require the use of ear protectors on passengers. Unlike motorboats, the sound generators on these vehicles are airborne, resulting in more noise impact. Florida requires a maximum level of 90 DB(A). Maine has three levels: operating, operating test and stationary test

Engine Braking Devices

This provision is an emission control. It can restrict brake use for only safety purposes and by defining restricted areas. It can require that mufflers be maintained to keep emitted sound to that of the original equipment. Common terminology is Jake brakes after the Jacobs Company. Milwaukee, Wi prohibits use within city limits. Portland, OR prohibits use within 200 feet of a residence. Albuquerque, NM posts signs requiring proper mufflers.

Motorboats

This provision is both an objective emission control and a subjective immission control. Because they are moving sources, objective controls are appropriate for measurements on open waterways. Many motorboats operate in bounding areas, such as small lakes or canals, with adjacent residential areas. In this case, immission controls are appropriate. California, Portland OR, and Seattle, WA require immission measurements to be made at the shoreline. Many states require emission measurements to be made at 50 feet.

Motor Vehicle Horns or Signaling Devices

This provision is both a prohibition and an emission control. it limits to use for safety warning only. It limits the sound level to a specific level at a prescribed distance. This provision is intended to limit horn use to safety and to limit the use of excessively loud air horns or Rumbler or Howler horns. California prohibits a person operating a motor vehicle to wear a headset or earplugs on both ears. Oregon prohibits signaling sound when an emergency vehicle is stationary or returning from an emergency

Motor Vehicles on a Public Right-of-Way

This provision is an objective emission control. It applies maximum sound levels to various categories of moving vehicles and for several vehicle speeds. It is the backbone of vehicle sound emission regulations. It generally requires a measurement of A-weighted sound level of a moving vehicle at a specific distance from the vehicle path (normally 50 feet). This provision has level restrictions on trucks over 10,000 GVW used locally and in interstate commerce. It also covers motorcycles of two horsepower ratings, mopeds, and all other vehicles on public rights-of-way. The federal government has set maximum levels for heavy trucks used in interstate commerce (40 CFR 202)and for motorcycles (40 CFR 205). Most states and many cities have maximum limits and they generally agree with federal standards where they apply. The most common speed division is 35 mph.

Motor Vehicle Racing Events

This provision is an objective emission control. It can define the method of vehicle operation that is used to define the maximum permitted sound level. It can have a curfew. Some states exempt motor vehicle racing events from noise disturbance litigation or prosecution. Arizona exempts racing motorcycles from maximum sound levels and muffler requirements. Illinois had detailed regulations on racing vehicles It required a 14 dB reduction in sound output, limited sound output at half mater to 115 dB(A), and no more than 105 dB(A) at 50 feet.

Motor Vehicle Sound Systems

This provision is both an objective emission control, a subjective emission control, and a subjective immission for vehicles on a public right-of-way. The first part limits the system sound level at a fixed distance. The second part uses the plainly audible definition for limiting the sound output. The third part uses the noise disturbance definition to limit the impact on neighboring properties and can be applied within public transportation. The most restrictive application of the plainly audible laws says that the sound cannot be audible to anyone other than the vehicle occupants. There are numerous state and community restrictions on vehicle sound systems. Louisiana prohibits the system from emitting sound outside of a vehicle. Richmond, CA also prohibits the sound from being audible outside the vehicle. Oregon prohibits sound systems being plainly audible at 50 feet. California prohibits sound systems that can be heard at 50 feet. Colorado Springs, CO, requires a measurement at 25 feet beyond the private property line or 25 feet from the source on public property; it does not specify a limiting level. In Lakewood, CO it must not be plainly audible beyond 25 feet. In Los Angeles, CA, it cannot be audible beyond 200 feet. In Seattle, WA, it must not be plainly audible at 75 feet. Chicago restricts levels to less than clearly audible at 75 feet. Minneapolis, MN restricts levels to less than audible at 50 feet. Albuquerque, NM restricts plainly audible to 25 feet, but also applies their land use limits. Cincinnati, OH restricts plainly audible to 50 feet. Dallas, TX prohibits sound or vibration that is detectable at 30 feet, or that violates the land use regulations. Houston, TX applies land use restrictions. Omaha, NE states the sound must not be audible at 100 feet. Hammond, IN restricts plainly audible to 25 feet. New Jersey states the sound must not be plainly audible at 50 feet between 8 am and 10 pm and not plainly audible at 25 feet between 10 pm and 8 am. Florida states the sound must not be plainly audible at 25 feet, but exempts business and political systems. Oregon and Tennessee state that the sound must not be plainly audible beyond 50 feet, as does Fairbanks, AK. Rhode Island specifically addresses low frequency sound that can be heard 20 feet from a closed vehicle or 100 feet otherwise. Salt Lake County Health Department, UT considers the sound a violation if it is plainly audible on a common carrier. Austin, TX states it must not be audible at 30 feet.

Motor Vehicle Theft Alarms

This provision is a subjective emission control with only an operational time limit. Los Angeles, CA () requires silencing in 5 minutes. New York City, NY requires automatic shut-off after 10 minutes and a prominent display of the local precinct number and telephone number. Boston, MA considers it a violation if the alarm is plainly audible at 200 feet and is on more than 5 minutes. Other states an communities have automatic shutoff times from 10 to 15 minutes. Some communities have banned such alarms.

Motor Vehicle Tire Squeal and Street Drag Racing

This provision is a subjective immission control. It is based on the noise disturbance from drag racing and tire squealing on public rights-of-way. Illinois (625 ILCS 5/11-505) prohibits such activities. Hammond, IN prohibits such activity if it creates a noise disturbance.

Railroads

Railroad activity is subject to federal regulations. Most communities do not attempt to regulate train sound. The level of train horns permitted by the Federal Railroad Administration sufficiently high that community impact occurs. One method to alleviate this sound is to have a community establish a quiet zone where the rail crossings meet federal safety standards so that horn use is not needed.

Recreational Off-Road Vehicles

This provision is both an objective emission control and a subjective immission control with a curfew. It can apply to both public and private properties. Since these vehicles can move in large open areas, an objective control, limiting maximum sound levels at a fixed distance, is appropriate. Since they also can move in bounded areas near residences, a subjective control is appropriate. Numerous states and cities have emission controls measured at 50 feet; the most common level is 82 dB(A), which is similar to that for motor vehicles on public rights-of-way. Colorado Springs, CO requires a minimum distance of 660 feet from residences. Portland, OR requires the area must be designated for recreational vehicle use. Salt Lake health Department, UT requires off-highway vehicles to be at least 800 feet from a dwelling during the day and has a curfew from 10 pm to 7 am. They prohibit any noise disturbance and require sound levels to the less than 96 dB(A) at 50 feet.

Recreational Snowmobiles

This provision is both an objective emission control and a subjective immission control with a curfew. It can apply to both public and private properties. Since these vehicles can move in large open areas, an objective control, limiting maximum sound levels at a fixed distance is appropriate. Since they also can move in bounded areas near residences, a subjective control is appropriate. Numerous states and communities have objective controls; the most common maximum level is 78 dB(A). Federal law (36 CFR 2.18) regulates snowmobiles on federal property at 78 dB(A), so states and communities are free to regulate snowmobile sound levels on their property. Lincoln NE limits levels to 78 dB(A). Maine exempts snowmobiles at sanctioned racing events.

Refuse Collection Vehicles

This provision is an objective emission control and a subjective control. It can place a maximum sound level at a specific distance for the loudest operation. It can set a curfew, or it can be based on noise disturbance in residential zones. Los Angeles, CA has a time limit that applies only within 200 feet of any residential building. Chicago, IL considers any noise it a noise disturbance if the activity occurs between 8pm to 8am. Salt Lake City, UT considers it a noise disturbance if the activities occur between 10 pm to 7 am and closer than 800 feet from a dwelling. Atlanta, GA prohibits collection between 9pm and am on a weekend day or legal holiday, except by permit. In Maryland, refuse collection is exempt during daytime hours and must meet maximum land use levels [55 dB(A)] in residential zones at night.

Standing Motor Vehicles

This provision is a subjective immission control. It sets a time limit on engine activity. It can also place a curfew on any engine activity. e. In Salt Lake City, UT it is considered a noise disturbance if the operation lasts more than 15 minutes. Dallas, TX applies the code to vehicles over 14,000 GVWR; they must be more than 300 feet from a residential zone and there is a 10-minute maximum. They also provide a list of idling vehicles that are exempt from prosecution such as buses or active concrete trucks. Hammond, IN limits operation to 3 minutes in an hour for vehicles over 14,000 GVWR in either public or private property. It exempts buses and taxis. Massachusetts allows idling no more than 5 minutes.

Building Codes

In the case of construction of new (or remodeled) apartments, condominiums, hospitals and hotels, many U.S. states and cities have stringent building codes with requirements of acoustical analysis, in order to protect building occupants from exterior noise sources and sound generated within the building itself. With regard to exterior noise, the codes usually require measurement of the exterior acoustic environment in order to determine the performance standard required for exterior building skin design.

The architect can work with the acoustical scientist to arrive at the best cost-effective means of creating a quiet interior (normally 45 dB). The most important elements of design of the building skin are usually: glazing (glass thickness, double pane design, etc.), roof material, caulking standards, chimney baffles, exterior door design, mail slots, attic ventilation ports and mounting of through the wall air conditioners. A special case of building skin design arises in the case of aircraft noise, where the FAA has funded extensive work in residential retrofit.

Regarding sound generated inside the building, there are two principal types of transmission. First, airborne sound travels through walls or floor/ceiling assemblies and can emanate from either human activities in adjacent living spaces or from mechanical noise within the building systems. Human activities might include voice, amplified sound systems or animal noise. Mechanical systems are elevator systems, boilers, refrigeration or air conditioning systems, generators and trash compactors. Since many of these sounds are inherently loud, the principle of regulation is to require the wall or ceiling assembly to meet certain performance standards (typically Sound Transmission Class of 50), which allows considerable attenuation of the sound level reaching occupants.

The second type of interior sound is called Impact Insulation Class (IIC) transmission. This effect arises not from airborne transmission, but rather from transmission of sound through the building itself. The most common perception of IIC noise is from footfall of occupants in living spaces above. This type of noise is somewhat more difficult to abate, but consideration must be given to isolating the floor assembly above or hanging the lower ceiling on resilient channel. Commonly a performance standard of IIC equal to 50 is specified in building codes. California has generally led the U.S. in widespread application of building code requirements for sound transmission; accordingly, the level of protection for building occupants has increased markedly in the last several decades.

References

- Noise-pollution, science: britannica.com, Retrieved 19 April, 2019
- "Safety and Health Topics | Occupational Noise Exposure | Occupational Safety and Health Administration". Www.osha.gov. Retrieved 2016-11-08
- Katz J, Chasin M, English KM, Hood LJ, Tillery KL (2014-09-19). Handbook of clinical audiology (Seventh ed.). Philadelphia. ISBN 978-1-4511-9163-9. OCLC 877024342s
- Levine RA, Oron Y (2015). Tinnitus. Handbook of Clinical Neurology. 129. Pp. 409–31. Doi:10.1016/B978-0-444-62630-1.00023-8. ISBN 9780444626301. PMID 25726282
- "Aircraft Airframe Noise—Research Overview". Archived from the original on 2008-05-17. Retrieved 2008-07-13
- Davies H, Kamp IV (2012-11-01). "Noise and cardiovascular disease: a review of the literature 2008-2011". Noise & Health. 14 (61): 287–91. Doi:10.4103/1463-1741.104895. PMID 23257579
- Schmid, RE (2007-02-18). "Aging nation faces growing hearing loss". CBS News. Archived from the original on November 15, 2007. Retrieved 2007-02-18
- How-to-control-noise-pollution-6-effective-measures, noise-pollution: yourarticlelibrary.com, Retrieved 10 July, 2019

6

Other Types of Environmental Pollution

Light pollution, radioactive contamination, visual pollution, particulate pollution, nanopollution, blackwater and drug pollution are some of the other kinds of environmental pollution. This chapter closely examines these types of environmental pollution to provide a thorough understanding of the subject.

Light Pollution

Light pollution is an unwanted or excessive artificial light. Like noise pollution, light pollution is a form of waste energy that can cause adverse effects and degrade environmental quality. Moreover, because light (transmitted as electromagnetic waves) is typically generated by electricity, which itself is usually generated by the combustion of fossil fuels, it can be said that there is a connection between light pollution and air pollution (from fossil-fueled power plant emissions). Control of light pollution therefore will help to conserve fuel (and money) and reduce air pollution as well as mitigate the more immediate problems caused by the excessive light. Although light pollution may not appear to be as harmful to public health and welfare as pollution of water resources or the atmosphere, it is an environmental quality issue of no small significance.

Light pollution.

Light pollution adversely affects professional and amateur astronomers, as well as casual observers of the night sky, because it severely reduces the visibility of stars and other celestial objects. The reduction in night sky visibility is a result of "skyglow," upward-directed light emanating from poorly designed or directed lamps and security floodlights. This wasted light is scattered and reflected by solid or liquid particles in the atmosphere and then returned to the eyes of people on the ground, obliterating their view of the night sky. The effect of skyglow from a town or city is not necessarily localized; it can be observed far from the main source.

Light pollution is a problem not only for astronomers and people who simply want to enjoy the beauty of a starry night. Glare from road lamps, commercial security lights and signs, or even from a neighbour's bright and misdirected yard lighting can cause discomfort and distraction and adversely affect the quality of life of many people. Light pollution also has adverse impacts on birds and other animals. Many migratory birds, for example, fly by night, when light from the stars and Moon helps them navigate. These birds are disoriented by the glare of artificial light as they fly over urban and suburban areas. It has been estimated by the American Bird Conservatory that more than four million migratory birds perish each year in the United States by colliding with brightly illuminated towers and buildings. Light pollution is considered to be one of the contributing factors in the dramatic decline of certain migratory songbird populations over the past several decades.

The quantity of light pollution from a given area depends on the number and brightness of light sources on the ground, the fraction of light that escapes above the horizontal, the reflectivity of surfaces near the light sources (e.g., roads, pavements, walls, windows), and the prevailing atmospheric conditions. Empirical formulas allow the calculation of skyglow levels as a function of population and distance from the observer. When skyglow levels are more than 10 percent above the natural background levels, significant sky degradation has begun. Even lights from a fairly small town with a population of only 3,000 people will cause significant night sky degradation for an observer as far as 10 km (6 miles) away.

Light pollution can be reduced by using well-designed light fixtures with modern optical controls to direct the light downward and also by using the minimum amount of wattage for the area to be illuminated. National and local government agencies can help by passing and enforcing appropriate light-control laws and ordinances.

Over Illumination

Over illumination is the presence of lighting intensity higher than that which is appropriate for a specific activity. Over illumination was commonly ignored between 1950 and 1995, especially in office and retail environments. Since then, however, the interior design community has begun to reconsider this practice. Over illumination encompasses two separate concerns:

- Unnecessary electric lighting is expensive and energy-intensive. Lighting accounts for approximately 9% of residential electricity use as of 2001 and about 40% of commercial electricity use.
- Excessive levels of artificial light may adversely affect health. These detrimental effects may depend on the spectrum as well as the overall brightness level of light.

Over illumination can be reduced by installing occupancy sensors, using natural sunlight whenever possible, turning off lights when leaving a room, or changing the type of lightbulb. Over illumination does not refer to snowblindness, where high exposure to ultraviolet light causes physical damage to the eye. Too little light, the opposite of over illumination, is associated with seasonal affective disorder.

Causes

Some big-box retail stores are over illuminated.

A fitness club that is illuminated mostly by natural light.

Over illumination can be caused by several factors:

- Illuminating an unoccupied area.
- Using electrical lights instead of natural light.
- Providing lighting for an occupied area, but with too much intensity.
- Installing too few electrical controls. This results in an area that must either be over illuminated or not illuminated at all.

In addition, there are ancillary reasons why over illumination might be decided to be necessary. For example, retail stores with large windows will sometimes stay illuminated overnight as a method of crime prevention.

While some aspects of illumination are readily controllable, such as turning off lights when exiting a room, others are determined by the architecture and construction of the building. For example, skylights decrease the amount of artificial lighting that is required during the daytime, but most buildings do not have them installed. In addition, too few light switches can also cause issues. If an office building with large windows only has one switch per floor, then electric lights will illuminate the perimeter areas (with abundant natural light) to same level as the interior zones (which receive less sunlight).

Health Effects

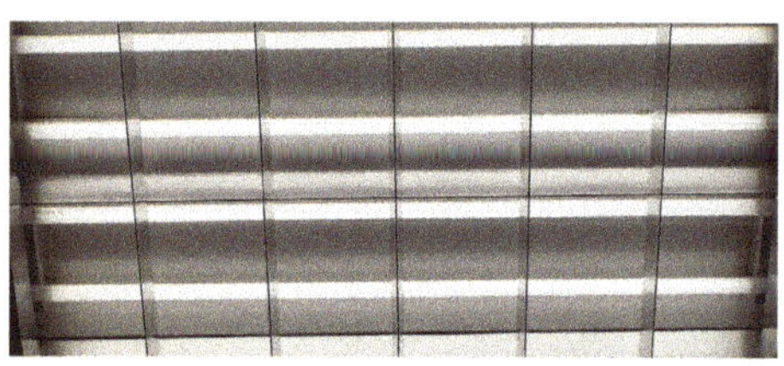

Common T8 fluorescent lighting fixture frequently used in office environments.

Over illumination has been linked to various negative health effects. While some effects might happen because the color spectrum of fluorescent lighting is significantly different from sunlight, other symptoms might be caused by light that is simply too intense. In particular, over illumination has been linked to headaches, fatigue, medically defined stress, anxiety, and decreases in sexual function.

Some studies attribute migraine headaches to overly intense light, while others linked it with certain spectral distributions. In one survey bright light was the number two trigger (affecting 47% of respondents) for causing a migraine episode.

Fatigue is another common complaint from individuals exposed to over illumination, especially with fluorescent media.

Likewise, over illumination can also cause stress and anxiety. In fact, natural light was preferred over purely artificial light by office workers from both eastern and western cultures. In addition, over illumination can cause medical stress and even aggravate other psychological disorders like agoraphobia. The replacement of natural light with artificial light also decreases task performance under certain conditions.

Circulatory and Circadian Rhythm Effects

Hypertension effects of over illumination can lead to aggravation of cardiovascular disease and erectile dysfunction, which impacts are outcomes of long term cumulative exposure and associated systematic increases in blood pressure. The mechanism of this effect seems to be stress by related upregulation of adrenaline production akin to the fight-or-flight response.

Circadian rhythm disruption is primarily caused by the wrong timing of light in reference to the circadian phase. It can also be affected by too much light, too little light, or incorrect spectral composition of light. This effect is driven by stimulus (or lack of stimulus) to photosensitive ganglion cells in the retina. The "time of day", the circadian phase, is signalled to the pineal gland, the body's photometer, by the suprachiasmatic nucleus. Bright light in the evening or in the early morning shifts the phase of the production of melatonin. An out-of-sync melatonin rhythm can worsen cardiac arrhythmias and increase oxidized lipids in the ischemic heart. Melatonin also reduces superoxide production and myeloperoxide (an enzyme in neutrophils which produces hypochlorous acid) during ischemia-reperfusion.

Energy and Economic Considerations

Using too much lighting will lead to higher electricity consumption, and therefore higher electricity costs, for the owner of the building. The solution to this issue is multi-faceted because there are several causes of too much light.

High-Intensity Light

First, over illumination occurs when the intensity of lighting is too high for a given activity. For example, an office building might have many sets of fluorescent lights to keep the area illuminated after sundown. During the daytime, however, large windows might allow for abundant amounts of natural sunlight to enter the office. Therefore, keeping all of the fluorescent lights illuminated during the daytime might cause unnecessary costs and energy consumption.

Leaving Lights Turned On

Likewise, not switching off the lights when exiting a room will also cause higher energy consumption. Some people avoid frequently turning off CFL bulbs because they think that doing so will cause them to burn out prematurely. While this is true to a certain extent, the US Department of Energy recommends that 15 minutes is an appropriate time frame. If someone plans to leave the room for less than 15 minutes, then the light should be left on. If the room will be unoccupied for more than 15 minutes, then the light should be switched off. Another concern is that turning on a fluorescent bulb consumes large amounts of energy. While fluorescent bulbs do need more energy turn on, the amount of electricity consumed is equal to only a few seconds of normal operation.

Sometimes people do not turn off lights for other reasons, such as an office worker whose company actually pays for the electricity. In these cases building automation provides more control. These solutions provide centralized control of all lighting within a home or commercial building, allowing easy implementation of scheduling, occupancy control, daylight harvesting and more. Many systems also support demand response and will automatically dim or turn off lights to take advantage of DR incentives and cost savings. Many newer control systems are using wireless mesh open standards (such as ZigBee), which provides benefits including easier installation (no need to run control wires) and interoperability with other standards-based building control systems (e.g. security).

Architectural Design and the Type of Light Bulbs

Architectural design can also provide ways of reducing energy usage. There are technological aspects of window design where window angles can be calculated to minimize interior glare and reduce interior over illumination, while at the same time reducing solar heat loading and subsequent demand for air conditioning as energy conservation techniques. For the Dakin Building in Brisbane, California the angled window projections effectively provide permanent sunscreens, obviating interior blinds or shades.

Ideally, the design of a building would create multiple switches for the overhead lighting. Adjusting these settings would allow the optimal light intensity to be delivered, the most common version of this control being the "three-level switch", also called A/B switching. Much of the benefit of the excess illumination reduction comes from a better ratio of natural light to fluorescent light that can result from any of the above changes. Research has been conducted showing worker productivity gains in settings where each worker selects his or her own lighting level.

Lastly, the type of light bulbs that are installed has a significant effect on energy consumption. The efficiency of light sources vary greatly. Fluorescent lights produce several times as much light, for given power input, as incandescent lights do, and LEDs are continuing to improve beyond that. Shades vary in their absorption. Light colored ceilings, walls and other surfaces increase ambient light by reflecting.

Radioactive Contamination

Radioactive contamination, also called radiological contamination, is the deposition of, or presence of radioactive substances on surfaces or within solids, liquids or gases (including the human body), where their presence is unintended or undesirable.

Such contamination presents a hazard because of the radioactive decay of the contaminants, which emit harmful ionising radiation such as alpha particles or beta particles, gamma rays or neutrons. The degree of hazard is determined by the concentration of the contaminants, the energy of the radiation being emitted, the type of radiation, and the proximity of the contamination to organs of the body. It is important to be clear that the contamination gives rise to the radiation hazard, and the terms "radiation" and "contamination" are not interchangeable.

The sources of radioactive pollution can be classified into two groups: natural and man made. Following an atmospheric nuclear weapon discharge or a nuclear reactor containment breach, the air, soil, people, plants, and animals in the vicinity will become contaminated by nuclear fuel and fission products. A spilled vial of radioactive material like uranyl nitrate may contaminate the floor and any rags used to wipe up the spill. Cases of widespread radioactive contamination include the Bikini Atoll, the Rocky Flats Plant in Colorado, the Fukushima Daiichi nuclear disaster, the Chernobyl disaster, and the area around the Mayak facility in Russia.

Sources of Contamination

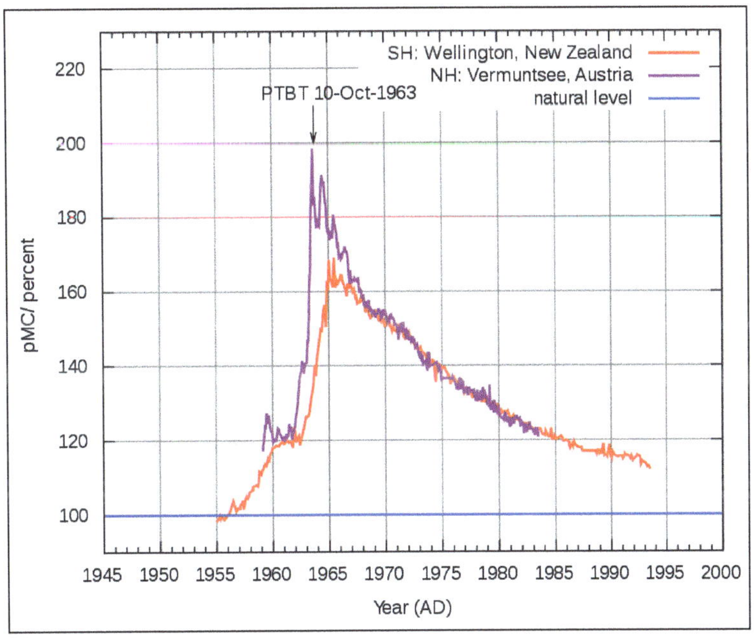

The sources of radioactive pollution can be natural or man-made. Radioactive contamination can be due to a variety of causes. It may occur due to release of radioactive gases, liquids or particles. For example, if a radionuclide used in nuclear medicine is spilled (accidentally or, as in the case of the Goiânia accident, through ignorance), the material could be spread by people as they walk around.

Radioactive contamination may also be an inevitable result of certain processes, such as the release of radioactive xenon in nuclear fuel reprocessing. In cases that radioactive material cannot be contained, it may be diluted to safe concentrations. For a discussion of environmental contamination by alpha emitters please see actinides in the environment.

Nuclear fallout is the distribution of radioactive contamination by the 520 atmospheric nuclear explosions that took place from the 1950s to the 1980s. In nuclear accidents, a measure of the type and amount of radioactivity released,such as from a reactor containment failure, is known as the source term. The United States Nuclear Regulatory Commission defines this as "Types and amounts of radioactive or hazardous material released to the environment following an accident."

Contamination does not include residual radioactive material remaining at a site after the completion of decommissioning. Therefore, radioactive material in sealed and designated containers is not properly referred to as contamination, although the units of measurement might be the same.

Containment

Large industrial glovebox in the nuclear industry.

Containment is the primary way of preventing contamination being released into the environment or coming into contact or being ingested by humans. Being within the intended Containment differentiates radioactive material from radioactive contamination. When radioactive materials are concentrated to a detectable level outside a containment, the area affected is generally referred to as "contaminated".

There are a large number of techniques for containing radioactive materials so that it does not spread beyond the containment and become contamination. In the case of liquids this is by the use of high integrity tanks or containers, usually with a sump system so that leakage can be detected by radiometric or conventional instrumentation.

Where material is likely to become airborne, then extensive use is made of the glovebox, which is a common technique in hazardous laboratory and process operations in many industries. The gloveboxes are kept under a slight negative pressure and the vent gas is filtered in high efficiency filters, which are monitored by radiological instrumentation to ensure they are functioning correctly.

Naturally Occurring Radioactivity

A variety of radionuclides occur naturally in the environment. Elements like uranium and thorium, and their decay products, are present in rock and soil. Potassium-40, a primordial nuclide, makes up a small percentage of all potassium and is present in the human body. Other nuclides, like carbon-14, which is present in all living organisms, are continuously created by cosmic rays.

These levels of radioactivity pose little danger but can confuse measurement. A particular problem is encountered with naturally generated radon gas which can affect instruments which are set to detect contamination close to normal background levels and can cause false alarms. Because of this skill is required by the operator of radiological survey equipment to differentiate between background radiation and the radiation which emanates from contamination.

Naturally occurring radioactive materials (NORM) can be brought to the surface or concentrated by human activities like mining, oil and gas extraction and coal consumption.

Control and Monitoring of Contamination

G-M counters being used as gamma survey monitors, seeking radioactive satellite debris.

Radioactive contamination may exist on surfaces or in volumes of material or air, and specialist techniques are used to measure the levels of contamination by detection of the emitted radiation.

Contamination Monitoring

Contamination monitoring depends entirely upon the correct and appropriate deployment and utilisation of radiation monitoring instruments.

Surface Contamination

Surface contamination may either be fixed or "free". In the case of fixed contamination, the radioactive material cannot by definition be spread, but its radiation is still measurable. In the case of free contamination there is the hazard of contamination spread to other surfaces such as skin or clothing, or entrainment in the air. A concrete surface contaminated by radioactivity can be shaved to a specific depth, removing the contaminated material for disposal.

For occupational workers controlled areas are established where there may be a contamination hazard. Access to such areas is controlled by a variety of barrier techniques, sometimes involving changes of clothing and foot wear as required. The contamination within a controlled area is normally regularly monitored. Radiological protection instrumentation (RPI) plays a key role in monitoring and detecting any potential contamination spread, and combinations of hand held survey instruments and permanently installed area monitors such as Airborne particulate monitors and area gamma monitors are often installed. Detection and measurement of surface contamination of personnel and plant is normally by Geiger counter, scintillation counter or proportional counter. Proportional counters and dual phosphor scintillation counters can discriminate between alpha and beta contamination, but the Geiger counter cannot. Scintillation detectors are generally preferred for hand held monitoring instruments, and are designed with a large detection window to make monitoring of large areas faster. Geiger detectors tend to have small windows, which are more suited to small areas of contamination.

Exit Monitoring

The spread of contamination by personnel exiting controlled areas in which nuclear material is used or processed is monitored by specialised installed exit control instruments such as frisk probes, hand contamination monitors and whole body exit monitors. These are used to check that persons exiting controlled areas do not carry contamination on their body or clothes.

In the United Kingdom the HSE has issued a user guidance note on selecting the correct portable radiation measurement instrument for the application concerned. This covers all radiation instrument technologies, and is a useful comparative guide for selecting the correct technology for the contamination type.

The UK NPL publishes a guide on the alarm levels to be used with instruments for checking personnel exiting controlled areas in which contamination may be encountered. Surface contamination is usually expressed in units of radioactivity per unit of area for alpha or beta emitters. For SI, this is becquerels per square meter (or Bq/m^2). Other units such as picoCuries per 100 cm^2 or disintegrations per minute per square centimeter (1 dpm/cm^2 = 167 Bq/m^2) may be used.

Airborne Contamination

The air can be contaminated with radioactive isotopes in particulate form, which poses a particular inhalation hazard. Respirators with suitable air filters, or completely self-contained suits with their own air supply can mitigate these dangers.

Airborne contamination is measured by specialist radiological instruments that continuously pump the sampled air through a filter. Airborne particles accumulate on the filter and can be measured in a number of ways:

- The filter paper is periodically manually removed to an instrument such as a "scaler" which measures any accumulated radioactivity.
- The filter paper is static and is measured in situ by a radiation detector.
- The filter is a slowly moving strip and is measured by a radiation detector. These are commonly called "moving filter" devices and automatically advance the filter to present a clean area for accumulation, and thereby allow a plot of airborne concentration over time.

Commonly a semiconductor radiation detection sensor is used that can also provide spectrographic information on the contamination being collected.

A particular problem with airborne contamination monitors designed to detect alpha particles is that naturally occurring radon can be quite prevalent and may appear as contamination when low contamination levels are being sought. Modern instruments consequently have "radon compensation" to overcome this effect.

Internal Human Contamination

Radioactive contamination can enter the body through ingestion, inhalation, absorption, or injection. This will result in a committed dose of radiation.

For this reason, it is important to use personal protective equipment when working with radioactive materials. Radioactive contamination may also be ingested as the result of eating contaminated plants and animals or drinking contaminated water or milk from exposed animals. Following a major contamination incident, all potential pathways of internal exposure should be considered. Successfully used on Harold McCluskey, chelation therapy and other treatments exist for internal radionuclide contamination.

Decontamination

Cleaning up contamination results in radioactive waste unless the radioactive material can be returned to commercial use by reprocessing. In some cases of large areas of contamination, the contamination may be mitigated by burying and covering the contaminated substances with concrete, soil, or rock to prevent further spread of the contamination to the environment. If a person's body is contaminated by ingestion or by injury and standard cleaning cannot reduce the contamination further, then the person may be permanently contaminated.

Contamination control products have been used by the U.S. Department of Energy (DOE) and the commercial nuclear industry for decades to minimize contamination on radioactive equipment and surfaces and fix contamination in place. "Contamination control products" is a broad term that includes fixatives, strippable coatings, and decontamination gels. A fixative product functions as a permanent coating to stabilize residual loose/transferable radioactive contamination by fixing it in place; this aids in preventing the spread of contamination and reduces the possibility of the contamination becoming airborne, reducing workforce exposure and facilitating future deactivation and decommissioning (D&D) activities. Strippable coating products are loosely adhered paint-like films and are used for their decontamination abilities. They are applied to surfaces with loose/transferable radioactive contamination and then, once dried, are peeled off, which removes the loose/transferable contamination along with the product. The residual radioactive contamination on the surface is significantly reduced once the strippable coating is removed. Modern strippable coatings show high decontamination efficiency and can rival traditional mechanical and chemical decontamination methods. Decontamination gels work in much the same way as other strippable coatings. The results obtained through the use of contamination control products is variable and depends on the type of substrate, the selected contamination control product, the contaminants, and the environmental conditions (e.g., temperature, humidity, etc.).

Some of the largest areas committed to be decontaminated are in the Fukushima Prefecture, Japan. The national government is under pressure to clean up radioactivity due to the Fukushima nuclear accident of March 2011 from as much land as possible so that some of the 110,000 displaced people can return. Stripping out the key radioisotope threatening health (caesium-137) from low level waste could also dramatically decrease the volume of waste requiring special disposal. A goal is to find techniques that might be able to strip out 80 to 95% of the caesium from contaminated soil and other materials, efficiently and without destroying the organic content in the soil. One being investigated is termed hydrothermal blasting. The caesium is broken away from soil particles and then precipitated with ferric ferricyanide (Prussian blue). It would be the only component of the waste requiring special burial sites. The aim is to get annual exposure from the contaminated environment down to one millisievert (mSv) above background. The most contaminated area where radiation doses are greater than 50 mSv/year must remain off limits, but some areas that are currently less than 5 mSv/year may be decontaminated allowing 22,000 residents to return.

To help with protection of people living in geographical areas which have been radioactively contaminated the International Commission on Radiological Protection has published a guide: "Publication 111 – Application of the Commission's Recommendations to the Protection of People Living in Long-term Contaminated Areas after a Nuclear Accident or a Radiation Emergency".

Contamination Hazards

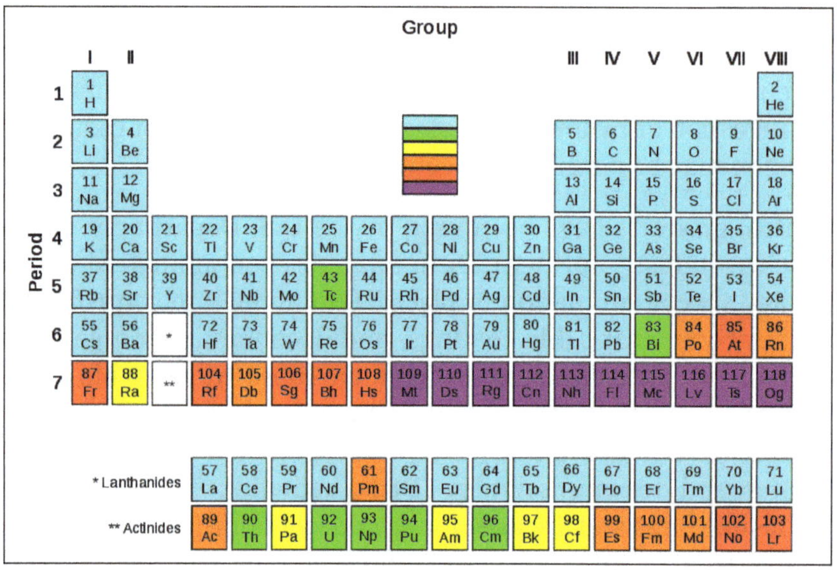

Periodic table with elements colored according to the half-life of their most stable isotope. Elements which contain at least one stable isotope.

- Radioactive elements: The most stable isotope is very long-lived, with half-life of over four million years.
- Radioactive elements: The most stable isotope has half-life between 800 and 34.000 years.
- Radioactive elements: The most stable isotope has half-life between one day and 130 years.
- Highly radioactive elements: The most stable isotope has half-life between several minutes and one day.
- Extremely radioactive elements: The most stable isotope has half-life less than several minutes.

Low-Level Contamination

The hazards to people and the environment from radioactive contamination depend on the nature of the radioactive contaminant, the level of contamination, and the extent of the spread of contamination. Low levels of radioactive contamination pose little risk, but can still be detected by radiation instrumentation. If a survey or map is made of a contaminated area, random sampling locations may be labeled with their activity in becquerels or curies on contact. Low levels may be reported in counts per minute using a scintillation counter.

In the case of low-level contamination by isotopes with a short half-life, the best course of action may be to simply allow the material to naturally decay. Longer-lived isotopes should be cleaned up and properly disposed of, because even a very low level of radiation can be life-threatening when in long exposure to it.

Facilities and physical locations that are deemed to be contaminated may be cordoned off by a health physicist and labeled "Contaminated area." Persons coming near such an area would typically require anti-contamination clothing ("anti-Cs").

High-Level Contamination

High levels of contamination may pose major risks to people and the environment. People can be exposed to potentially lethal radiation levels, both externally and internally, from the spread of contamination following an accident (or a deliberate initiation) involving large quantities of radioactive material. The biological effects of external exposure to radioactive contamination are generally the same as those from an external radiation source not involving radioactive materials, such as x-ray machines, and are dependent on the absorbed dose.

When radioactive contamination is being measured or mapped in situ, any location that appears to be a point source of radiation is likely to be heavily contaminated. A highly contaminated location is colloquially referred to as a "hot spot." On a map of a contaminated place, hot spots may be labeled with their "on contact" dose rate in mSv/h. In a contaminated facility, hot spots may be marked with a sign, shielded with bags of lead shot, or cordoned off with warning tape containing the radioactive trefoil symbol.

The radiation warning symbol (trefoil).

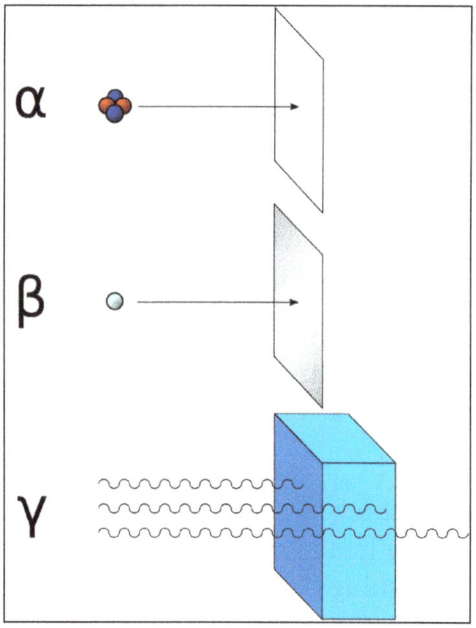

Alpha radiation consists of helium-4 nucleus and is readily stopped by a sheet of paper. Beta radiation, consisting of electrons, is halted by an aluminium plate. Gamma radiation is eventually absorbed as it penetrates a dense material. Lead is good at absorbing gamma radiation, due to its density.

The hazard from contamination is the emission of ionising radiation. The principal radiations which will be encountered are alpha, beta and gamma, but these have quite different characteristics. They have widely differing penetrating powers and radiation effect, and the accompanying diagram shows the penetration of these radiations in simple terms.

Radiation monitoring involves the measurement of radiation dose or radionuclide contamination for reasons related to the assessment or control of exposure to radiation or radioactive substances, and the interpretation of the results. The methodological and technical details of the design and operation of environmental radiation monitoring programmes and systems for different radionuclides, environmental media and types of facility are given in IAEA Safety Standards Series No. RS–G-1.8 and in IAEA Safety Reports Series No. 64.

Health Effects of Contamination

Biological Effects

Radioactive contamination by definition emits ionizing radiation, which can irradiate the human body from an external or internal origin.

External Irradiation

This is due to radiation from contamination located outside the human body. The source can be in the vicinity of the body or can be on the skin surface. The level of health risk is dependent on duration and the type and strength of irradiation. Penetrating radiation such as gamma rays, X-rays, neutrons or beta particles pose the greatest risk from an external source. Low penetrating radiation such as alpha particles have a low external risk due to the shielding effect of the top layers of skin.

Internal Irradiation

Radioactive contamination can be ingested into the human body if it is airborne or is taken in as contamination of food or drink, and will irradiate the body internally. The art and science of assessing internally generated radiation dose is Internal dosimetry.

The biological effects of ingested radionuclides depend greatly on the activity, the biodistribution, and the removal rates of the radionuclide, which in turn depends on its chemical form, the particle size, and route of entry. Effects may also depend on the chemical toxicity of the deposited material, independent of its radioactivity. Some radionuclides may be generally distributed throughout the body and rapidly removed, as is the case with tritiated water.

Some organs concentrate certain elements and hence radionuclide variants of those elements. This action may lead to much lower removal rates. For instance, the thyroid gland takes up a large percentage of any iodine that enters the body. Large quantities of inhaled or ingested radioactive iodine may impair or destroy the thyroid, while other tissues are affected to a lesser extent. Radioactive iodine-131 is a common fission product; it was a major component of the radioactivity released from the Chernobyl disaster, leading to nine fatal cases of pediatric thyroid cancer and hypothyroidism. On the other hand, radioactive iodine is used in the diagnosis and treatment of many diseases of the thyroid precisely because of the thyroid's selective uptake of iodine.

The radiation risk proposed by the International Commission on Radiological Protection (ICRP) predicts that an effective dose of one sievert (100 rem) carries a 5.5% chance of developing cancer. Such a risk is the sum of both internal and external radiation dose.

The ICRP states "Radionuclides incorporated in the human body irradiate the tissues over time periods determined by their physical half-life and their biological retention within the body. Thus they may give rise to doses to body tissues for many months or years after the intake. The need to regulate exposures to radionuclides and the accumulation of radiation dose over extended periods of time has led to the definition of committed dose quantities". The ICRP further states "For internal exposure, committed effective doses are generally determined from an assessment of the intakes of radionuclides from bioassay measurements or other quantities (e.g., activity retained in the body or in daily excreta). The radiation dose is determined from the intake using recommended dose coefficients".

The ICRP defines two dose quantities for individual committed dose:

- Committed equivalent dose, H T(t) is the time integral of the equivalent dose rate in a particular tissue or organ that will be received by an individual following intake of radioactive material into the body by a Reference Person, where t is the integration time in years. This refers specifically to the dose in a specific tissue or organ, in a similar way to external equivalent dose.
- Committed effective dose, E(t) is the sum of the products of the committed organ or tissue equivalent doses and the appropriate tissue weighting factors WT, where t is the integration time in years following the intake. The commitment period is taken to be 50 years for adults, and to age 70 years for children. This refers specifically to the dose to the whole body, in a similar way to external effective dose.

Social and Psychological Effects

A 2015 report in Lancet explained that serious impacts of nuclear accidents were often not directly attributable to radiation exposure, but rather social and psychological effects. The consequences of low-level radiation are often more psychological than radiological. Because damage from very-low-level radiation cannot be detected, people exposed to it are left in anguished uncertainty about what will happen to them. Many believe they have been fundamentally contaminated for life and may refuse to have children for fear of birth defects. They may be shunned by others in their community who fear a sort of mysterious contagion.

Forced evacuation from a radiological or nuclear accident may lead to social isolation, anxiety, depression, psychosomatic medical problems, reckless behavior, even suicide. Such was the outcome of the 1986 Chernobyl nuclear disaster in Ukraine. A comprehensive 2005 study concluded that "the mental health impact of Chernobyl is the largest public health problem unleashed by the accident to date". Frank N. von Hippel, a U.S. scientist, commented on 2011 Fukushima nuclear disaster, saying that "fear of ionizing radiation could have long-term psychological effects on a large portion of the population in the contaminated areas". Evacuation and long-term displacement of affected populations create problems for many people, especially the elderly and hospital patients.

Such great psychological danger does not accompany other materials that put people at risk of cancer and other deadly illness. Visceral fear is not widely aroused by, for example, the daily emissions from coal burning, although, as a National Academy of Sciences study found, this causes 10,000 premature deaths a year in the US population of 317,413,000. Medical errors leading to death in U.S. hospitals are estimated to be between 44,000 and 98,000. It is "only nuclear radiation that bears a huge psychological burden – for it carries a unique historical legacy".

Blackwater

Blackwater is a form of pollution produced in coal preparation. In its purification, coal is crushed in a coal preparation plant and then separated and transported as a coal slurry, From the slurry, incombustible materials are removed and the coal can be sized. After the recovery of the coal particles from this slurry, the remaining water is black, contains very fine particles of coal. This blackwater cannot be processed in a water treatment plant.

Disasters

Impoundments for storage of blackwater and other coal-related wastes have a troubled history with often severe environmental consequences.

In February 1972, three dams holding a mixture of coal slurry in Logan County, West Virginia failed in succession: 130,000,000 US gallons (490,000 m3) of toxic water were released in the Buffalo Creek Flood. As discussed in the book The Buffalo Creek Flood: An Act of Man, out of a population of 5,000 people, 125 people were killed, 1,121 were injured, and over 4,000 were left homeless. The flood caused 50 million dollars in damages. Despite evidence of negligence, the Pittston Coal Company, which owned the compromised dam, called the event an "Act of God." In 2002, a 900-foot (270 m) high, 2,000-foot (610 m) long valley fill in Lyburn, West Virginia failed and slid into a sediment pond at the toe of the fill, generating a large wave of water and sediment that destroyed several cars and houses.

Other Slurry Disasters

- Buffalo Creek Flood.
- Aberfan Disaster.
- Martin County sludge spill.
- Kingston Fossil Plant coal fly ash slurry spill.
- Little Blue Run Lake.

Future Technologies

The ultimate solution to the blackwater problem is to process coal without the use of water. Such dry-separation technologies are under development.

Visual Pollution

Visual pollution is an aesthetic issue and refers to the impacts of pollution that impair one's ability to enjoy a vista or view.

Visual pollution disturbs the visual areas of people by creating harmful changes in the natural environment. Billboards, open storage of trash, antennas, electric wires, buildings, and automobiles are often considered visual pollution. An overcrowding of an area causes visual pollution. Visual pollution is defined as the whole of irregular formations, which are mostly found in nature.

Effects of exposure to visual pollution include: distraction, eye fatigue, decreases in opinion diversity, and loss of identity.

Sources

Local managers of urban areas sometimes lack control over what is built and assembled in public places. As businesses look for ways to increase the profits, cleanliness, architecture, logic and use of space in urban areas are suffering from visual clutter. Variations in the built environment are determined by the location of street furniture such as public transport stations, garbage cans, large panels and stalls. Insensitivity of local administration is another cause for visual pollution. For example, poorly planned buildings and transportation systems create visual pollution. High-rise buildings, if not planned properly or sufficiently, can bring adverse change to the visual and physical characteristics of a city, which may reduce said city's readability.

A frequent criticism of advertising is that there is too much of it. Billboards, for example, have been alleged to distract drivers, corrupt public taste, promote meaningless and wasteful consumerism and clutter the land. See highway beautification. However, with the introduction of new communication technologies the fragmentation and incentive nature of advertising methods will improve, reducing clutter. Thus, with the increase of mobile device usage, more money goes to advertising on social media websites and mobile apps. Vandalism, in the form of graffiti is defined as street markings, offensive, inappropriate, and tasteless messages made without the owner's consent. Graffiti adds to visual clutter as it disturbs the view.

Artificial tree to hide a mobile phone base station.

Prevention

In the United States has,there are several initiatives gradually taking place to prevent visual pollution. The Federal Highway Beautification Act of 1965 limits placement of billboards on Interstate highways and federally aided roads. It has dramatically reduced the amount of billboards placed on these roads. Another highway bill, the Intermodal Surface Transportation Efficiency Act (ISTEA) of 1991 has made transportation facilities sync with the needs of communities. This bill created a system of state and national scenic byways and provided funds for biking trails, historic preservation and scenic conservation.

The Dunn Foundation is an organization that increases public awareness of visual pollution and landscape appearance in America through educational programs. The foundation has designed an educationally interactive package for students from grades 3-12 on how to improve the visual environment in their communities. Another company working toward prevention of visual clutter is Scenic America; a non-profit organization that envisions a future movement toward ensuring that scenic conservation boosts the economy and decreases visual pollution. Businesses situated near an interstate can create problems of advertising through large billboards, however now an alternative solution for advertisers is gradually eliminating the problem. For example, logo signs that provide directional information for travelers without disfiguring the landscape are increasing and are a step toward decreasing visual pollution on highways in America. Thus, researchers believe that planners should help and encourage citizens to maintain their communities as citizens have the power to influence government, especially local and regional management where most issues regarding appearance and disclosed.

In September 2006, São Paulo passed the Cidade Limpa (Clean City Law), outlawing the use of all outdoor advertisements, including on billboards, transit, and in front of stores.

Nanopollution

Nanopollution is a generic name for waste generated by nanodevices or during the nanomaterials manufacturing process. Ecotoxicological impacts of nanoparticles and the potential for bioaccumulation in plants and microorganisms is a subject of current research, as nanoparticles are considered to present novel environmental impacts. Of the US$710 million spent in 2002 by the U.S. government on nanotechnology research, $500,000 was spent on environmental impact assessments.

Nanowaste is mainly the group of particles that are released into the environment, or the particles that are thrown away when still on their products. The thrown away nanoparticles are usually still functioning how they are supposed to (still have their individual properties), they are just not being properly used anymore. Most of the time, they are lost due to contact with different environments. Silver nanoparticles, for example, they are used a lot in clothes to control odor, those particles are lost when washing them. The fact that they are still functioning and are so small is what makes nanowaste a concern. It can float in the air and might easily penetrate animal and plant cells causing unknown effects. Due to its small size, nanoparticles can have different properties than their own material when on a bigger size, and they are also functioning more efficiently because of its greater surface area. Most human-made nanoparticles do not appear in nature, so living organisms may not have appropriate means to deal with nanowaste.

Tinkle et al opine that, the presence of nanomaterials in our surroundings is not a herald of possible environmental hazards, because, the propensity towards toxicity as far as nanomaterials are concerned revolves around certain properties such as: Size, Geometry, Surface properties (charge, surface chemistry), Crystal structure and Chemical composition. Hund-Rinke et al, have established that mundane climatic dispositions/parameters could affect the properties of nanomaterials and consequently alter their bioavailability, hence potential toxicity/hazard, such environmental parameters include salinity and pH value.

The capacity for nanoparticles to function as a transport mechanism also raises concern about the transport of heavy metals and other environmental contaminants. Two areas of concern can be identified. First, in their free form nanoparticles can be released into the air or water during production, or production accidents, or as waste by-product of production, and ultimately accumulate in the soil, water, or plant life. Second, in fixed form, where they are part of a manufactured substance or product, they will ultimately have to be recycled or disposed of as waste.

Scrinis raises concerns about nano-pollution, and argues that it is not currently possible to "precisely predict or control the ecological impacts of the release of these nano-products into the environment." A May 2007 Report to the UK Department for Environment, Food and Rural Affairs noted concerns about the toxicological impacts of nanoparticles in relation to both hazard and exposure. The report recommended comprehensive toxicological testing and independent performance tests of fuel additives. Risks have been identified by Uskokovic in 2007. Concerns have also been raised about Silver Nano technology used by Samsung in a range of appliances such as washing machines and air purifiers.

One known consequences of human exposure to metals is the development of illnesses such as argyria and argyrosis caused by chronic exposure to soluble silver compounds. Silver can also cause some environmental problems. Due to its antimicrobial (antibacterial) properties, when present in soil it can kill beneficial bacteria that are important to keep the soil healthy. Environmental assessment is justified as nanoparticles present novel environmental impacts. Scrinis raises concerns about nano-pollution, and argues that it is not currently possible to "precisely predict or control the ecological impacts of the release of these nano-products into the environment."

Metals, in particular, exhibit strong bonds in compounds down to the nanoscale. Metals can stay and damage the environment for a long time, since they hardly degrade or get destroyed. With the increase in use of nanotechnology, it is predicted that the nanowaste of metals will keep increasing, and until a solution is found for that problem, that waste will keep accumulating in the environment. On the other hand, some possible future applications of nanotechnology have the potential to benefit the environment. Nanofiltration, based on the use of membranes with extremely small pores smaller than 10 nm (perhaps composed of nanotubes) are suitable for a mechanical filtration for the removal of ions or the separation of different fluids. One study has found a solution to filtrate and extract those nanoparticles from water. The process is still under investigation, but simulations have shown the total of about 90% to 99% removal of nanowaste particles from the water at an upgraded waste water treatment plant. Once the particles are separated from the water, they go to the landfill with the rest of the solids. Furthermore, magnetic nanoparticles offer an effective and reliable method to remove heavy metal contaminants from wastewater. Using nanoscale particles increases the efficiency of contaminant absorption and is comparatively inexpensive to traditional precipitation and filtration methods. One current method to recover nanoparticles is cloud point extraction, which allows pre-concentrated gold or other metallic nanoparticles to be extracted from aqueous solutions by manipulating temperature and surfactant concentration.

Life Cycle Responsibility

To properly assess the health hazards of engineered nanoparticles the whole life cycle of these particles needs to be evaluated, including their fabrication, storage and distribution, application and potential abuse, and disposal. The impact on humans or the environment may vary at different stages of the life cycle.

The Royal Society report identified a risk of nanoparticles or nanotubes being released during disposal, destruction and recycling, and recommended that "manufacturers of products that fall under extended producer responsibility regimes such as end-of-life regulations publish procedures outlining how these materials will be managed to minimize possible human and environmental exposure". Reflecting the challenges for ensuring responsible life cycle regulation, the Institute for Food and Agricultural Standards has proposed standards for nanotechnology research and development should be integrated across consumer, worker and environmental standards. They also propose that NGOs and other citizen groups play a meaningful role in the development of these standards.

Particulate Pollution

Particulate pollution is pollution of an environment that consists of particles suspended in some medium. There are three primary forms: atmospheric particulate matter, marine debris, and space debris. Some particles are released directly from a specific source, while others form in chemical reactions in the atmosphere. Particulate pollution can be derived from either natural sources or anthropogenic processes.

Atmospheric Particulate Matter

Atmospheric particulate matter, also known as particulate matter, or PM, describes solids and/or liquid particles suspended in a gas, most commonly the Earth's atmosphere. Particles in the atmosphere can be divided into two types, depending on the way they are emitted. Primary particles, such as mineral dust, are emitted into the atmosphere. Secondary particles, such as ammonium nitrate, are formed in the atmosphere through gas-to-particle conversion.

Marine Debris

Marine debris and marine aerosols refer to particulates suspended in a liquid, usually water on the Earth's surface. Particulates in water are a kind of water pollution measured as total suspended solids, a water quality measurement listed as a conventional pollutant in the U.S. Clean Water Act, a water quality law. Notably, some of the same kinds of particles can be suspended both in air and water, and pollutants specifically may be carried in the air and deposited in water, or fall to the ground as acid rain. The majority of marine aerosols are created through the bubble bursting of breaking waves and capillary action on the ocean surface due to the stress exerted from surface winds. Among common marine aerosols, pure sea salt aerosols are the major component of marine aerosols with an annual global emission between 2,000-10,000 teragrams annually. Through interactions with water, many marine aerosols help to scatter light, and aid in cloud condensation and ice nuclei (IN); thus, affecting the atmospheric radiation budget. When they interact with anthropogenic pollution, marine aerosols can affect biogeochemical cycles through the depletion of acids such as nitric acid and halogens.

Space Debris

Space debris describes particulates in the vacuum of outer space, specifically particles originating from human activity that remain in geocentric orbit around the Earth. The International Association of Astronauts define space debris as “any man-made Earth orbiting object which is non-functional with no reasonable expectation of assuming or resuming its intended function or any other function for which it is or can be expected to be authorized, including fragments and parts thereof”.

Space debris is classified by size and operational purpose, and divided into four main subsets: inactive payloads, operational debris, fragmentation debris and microparticulate matter. Inactive payloads refer to any launched space objects that have lost the capability to reconnect to its corresponding space operator; thus, preventing a return to Earth. In contrast, operational debris describes the matter associated with the propulsion of a larger entity into space, which may include upper rocket stages and ejected nose cones. Fragmentation debris refers to any object in space that has become dissociated from a larger entity by means of explosion, collision or deterioration. Microparticulate matter describes space matter that typically cannot be seen singly with the naked eye, including particles, gases, and spaceglow.

In response to research that concluded that impacts from Earth orbital debris could lead to greater hazards to spacecraft than the natural meteoroid environment, NASA began the orbital debris program in 1979, initiated by the Space Sciences branch at Johnson Space Center (JSC). Beginning with an initial budget of $70,000, the NASA orbital debris program began with the initial goals of characterizing hazards induced by space debris and creating mitigation standards that would minimize the growth of the orbital debris environment. By 1990, the NASA orbital debris program created a debris monitoring program, which included mechanisms to sample the low Earth orbit (LEO) environment for debris as small as 6mm using the Haystack X-band ground radar.

Epidemiology

Particulate pollution is observed around the globe in varying sizes and compositions and is the focus of many epidemiological studies. Particulate matter (PM) is generally classified into two main size categories: PM_{10} and $PM_{2.5}$. PM10, also known as coarse particulate matter, consists of particles 10 micrometers (μm) and smaller, while $PM_{2.5}$, also called fine particulate matter, consists of particles 2.5 μm and smaller. Particles 2.5 μm or smaller in size are especially notable as they can be inhaled into the lower respiratory system, and with enough exposure, absorbed into the bloodstream. Particulate pollution can occur directly or indirectly from a number of sources including, but not limited to: agriculture, automobiles, construction, forest fires, chemical pollutants, and power plants.

Exposure to particulates of any size and composition may occur acutely over a short duration, or chronically over a long duration. Particulate exposure has been associated with adverse respiratory symptoms ranging from irritation of the airways, aggravated asthma, coughing, and difficulty breathing from acute exposure to symptoms such as irregular heartbeat, lung cancer, kidney disease, chronic bronchitis, and premature death in individuals who suffer from pre-existing cardiovascular or lung diseases due to chronic exposure. The severity of health effects generally depends upon the size of the particles as well as the health status of the individual exposed; older adults, children, pregnant women, and immunocompromised populations are at the greatest risk for adverse health outcomes.

As a result, the US Environmental Protection Agency (EPA) and various health agencies around the world have established thresholds for concentrations of $PM_{2.5}$ and PM10 that are determined to be acceptable. However, there is no known safe level of exposure and thus, any exposure to particulate pollution is likely to increase an individual's risk of adverse health effects. In European countries, air quality at or above 10 micrograms per cubic meter of air (μg/m³) for $PM_{2.5}$ increases the all-causes daily mortality rate by 0.2-0.6% and the cardiopulmonary mortality rate by 6-13%.

Worldwide, PM_{10} concentrations of 70μg/m³ and $PM_{2.5}$ concentrations of 35μg/m³ have been shown to increase long-term mortality by 15%. More so, approximately 4.2 million of all premature deaths observed in 2016 occurred due to airborne particulate pollution, 91% of which occurred in countries with low to middle socioeconomic status. Of these premature deaths, 58% were attributed to strokes and ischaemic heart diseases, 8% attributed to COPD (Chronic Obstructive Pulmonary Disease), and 6% to lung cancer.

In 2006, the EPA conducted air quality designations in all 50 states, denoting areas of high pollution based on criteria such as air quality monitoring data, recommendations submitted by the states, and other technical information; and reduced the National Ambient Air Quality Standard for daily exposure to particulates in the 2.5 micrometers and smaller category from 15μg/m³ to 12μg/m³ in 2012. As a result, U.S. annual $PM_{2.5}$ averages have decreased from 13.5μg/m³ to 8.02μg/m³, between 2000 and 2017.

Drug Pollution

Drug pollution or pharmaceutical pollution is pollution of the environment with pharmaceutical drugs and their metabolites, which reach the aquatic environment (groundwater, rivers, lakes, and oceans) through wastewater. Drug pollution is therefore mainly a form of water pollution.

"Pharmaceutical pollution is now detected in waters throughout the world," said a scientist at the Cary Institute of Ecosystem Studies in Millbrook, New York. "Causes include aging infrastructure, sewage overflows and agricultural runoff. Even when wastewater makes it to sewage treatment facilities, they aren't equipped to remove pharmaceuticals."

Sources and Effects

Most such pollution comes simply from the drugs having been cleared and excreted in the urine. The portion that comes from expired or unneeded drugs that are flushed unused down the toilet is smaller, but it is also important, especially in hospitals (where its magnitude is greater than in residential contexts). This include drug molecules that are too small to be filtered out by existing water treatment plants. The process of upgrading existing plants to use advanced oxidation processes that are able to remove these molecules can be expensive. Drugs such as antidepressants have been found in the United States Great Lakes. Researchers from the University of Buffalo have found high traces of antidepressants in the brains of fish. Fish behavior on antidepressants have been noted to have similar impacts and reducing risk-averse behavior, and thereby reducing survival through predation.

Other sources include agricultural runoff (because of antibiotic use in livestock) and pharmaceutical manufacturing. Drug pollution is implicated in the sex effects of water pollution. It is a suspected a contributor (besides industrial pollution) in fish kills, amphibian dieoffs, and amphibian pathomorphology.

Prevention

The main action for preventing drug pollution is to incinerate unwanted pharmaceutical drugs. Burning them chemically degrades their active molecules, with few exceptions. The resulting ash can be further processed before landfilling, such as to remove and recycle any heavy metals that may be present.

There are now programs in many cities that provide collection points at places including drug stores, grocery stores, and police stations. People can bring their unwanted pharmaceuticals there for safe disposal, instead of flushing them (externalizing them to the waterways) or throwing them in the trash (externalizing them to a landfill, where they can become leachate).

Another aspect of drug pollution prevention is environmental law and regulation, although this faces the problems of enforcement costs, enforcement corruption and negligence, where enforcement succeeds, increased costs of doing business. The lobbying of pros and cons is ongoing.

Manufacturing

One extreme example of drug pollution was found in India in 2009 in an area where pharmaceutical manufacturing activity is concentrated. Not all pharmaceutical manufacturing contributes to the problem. In places where environmental law and regulation are adequately enforced, the wastewater from the factories is cleaned to a safe level. But to the extent that the market rewards "looking the other way" in developing nations, whether through local corruption (bribed inspectors or regulators) or plausible deniability, such protections are circumvented. This problem belongs to everyone, because consumers in well-regulated places constitute the biggest customers of the factories that operate in the inadequately regulated or inspected places, meaning that externality is involved.

References

- Light-pollution, science: britannica.com, Retrieved 18 April, 2019
- Biological Effects of Power Frequency Electric and Magnetic Fields, Office of Technology Assessment, U.S. Congress, University Press of the Pacific (2002) ISBN 0-89875-974-9
- Omidvarborna; et al. (2015). "Recent studies on soot modeling for diesel combustion". Renewable and Sustainable Energy Reviews. 48: 635–647. Doi:10.1016/j.rser.2015.04.019
- International Atomic Energy Agency (2010). Programmes and Systems for Source and Environmental Radiation Monitoring. Safety Reports Series No. 64. Vienna: IAEA. P. 234. ISBN 978-92-0-112409-8

- Yilmaz, Demet (May 2011). "In the Context of Visual Pollution: Effects to Trabzon City Center Silhoutte". The Asian Social Science Journal. 7 (5): 99. Retrieved March 30, 2013
- Narisada Kohei and Duco Schreude, Light Pollution Handbook, Springer, Netherlands (2004) ISBN 1-4020-2665-X
- "Five Years After Banning Outdoor Ads, Brazil›s Largest City Is More Vibrant Than Ever". The Center for a New American Dream. Retrieved September 26, 2013
- D. Woodruff and L. Macnamara, "Treatment of coal tailings" in "The Coal Handbook: Towards Cleaner Production" D Osborne, Ed. Woodhead Publishing, 2013. Print ISBN 0-85709-422-X. Web ISBN 978-0-85709-730-9

Permissions

All chapters in this book are published with permission under the Creative Commons Attribution Share Alike License or equivalent. Every chapter published in this book has been scrutinized by our experts. Their significance has been extensively debated. The topics covered herein carry significant information for a comprehensive understanding. They may even be implemented as practical applications or may be referred to as a beginning point for further studies.

We would like to thank the editorial team for lending their expertise to make the book truly unique. They have played a crucial role in the development of this book. Without their invaluable contributions this book wouldn't have been possible. They have made vital efforts to compile up to date information on the varied aspects of this subject to make this book a valuable addition to the collection of many professionals and students.

This book was conceptualized with the vision of imparting up-to-date and integrated information in this field. To ensure the same, a matchless editorial board was set up. Every individual on the board went through rigorous rounds of assessment to prove their worth. After which they invested a large part of their time researching and compiling the most relevant data for our readers.

The editorial board has been involved in producing this book since its inception. They have spent rigorous hours researching and exploring the diverse topics which have resulted in the successful publishing of this book. They have passed on their knowledge of decades through this book. To expedite this challenging task, the publisher supported the team at every step. A small team of assistant editors was also appointed to further simplify the editing procedure and attain best results for the readers.

Apart from the editorial board, the designing team has also invested a significant amount of their time in understanding the subject and creating the most relevant covers. They scrutinized every image to scout for the most suitable representation of the subject and create an appropriate cover for the book.

The publishing team has been an ardent support to the editorial, designing and production team. Their endless efforts to recruit the best for this project, has resulted in the accomplishment of this book. They are a veteran in the field of academics and their pool of knowledge is as vast as their experience in printing. Their expertise and guidance has proved useful at every step. Their uncompromising quality standards have made this book an exceptional effort. Their encouragement from time to time has been an inspiration for everyone.

The publisher and the editorial board hope that this book will prove to be a valuable piece of knowledge for students, practitioners and scholars across the globe.

Index

S

T

U

V

W

CPSIA information can be obtained
at www.ICGtesting.com
Printed in the USA
LVHW062018100122
708212LV00002B/318

9 781641 166218